Advances in Veterinary Science and Comparative Medicine

Volume 38B

Comparative Vertebrate Exercise Physiology: Phyletic Adaptations

Advances in Veterinary Science and Comparative Medicine

Edited by

Charles E. Cornelius
Department of Animal Resources
Scripps Research Institute
La Jolla, California

R. R. Marshak
New Bolton Center
University of Pennsylvania
Kennett Square, Pennsylvania

E. C. Melby
SmithKline Beecham
Animal Health Products Research and Development
West Chester, Pennsylvania

Advances in Veterinary Science and Comparative Medicine

Volume 38B

Comparative Vertebrate Exercise Physiology: Phyletic Adaptations

Edited by

James H. Jones

Department of Surgical and Radiological Sciences
School of Veterinary Medicine
University of California, Davis
Davis, California

Academic Press

San Diego New York Boston
London Sydney Tokyo Toronto

This book is printed on acid-free paper. ♾

Academic Press, Inc.
A Division of Harcourt Brace & Company
525 B Street, Suite 1900, San Diego, California 92101-4495

United Kingdom Edition published by
Academic Press Limited
24-28 Oval Road, London NW1 7DX

International Standard Serial Number: 0065-3519

International Standard Book Number: 0-12-039239-9

PRINTED IN THE UNITED STATES OF AMERICA
94 95 96 97 98 99 EB 9 8 7 6 5 4 3 2 1

CONTENTS

Contributors vii
Preface ix

Exercise Performance of Fish

Paul W. Webb

I. Introduction 1
II. What Is Performance? 3
III. Locomotor Range Fractionation 6
IV. Classification of Performance Levels 12
V. Limits to Performance 20
VI. Conclusions 40
References 42

Jumping Ability of Anuran Amphibians

Richard L. Marsh

I. Introduction 51
II. Mechanics of Jumping 52
III. Providing the Power: The Muscles Used in Jumping 63
IV. Jumping Performance 78
V. Detailed Studies of Jumping 93
VI. Conclusions 103
VII. Abbreviations 105
References 106

Exercise Performance of Reptiles

Albert F. Bennett

I. Introduction 113

II. Endurance Activity 116
III. Burst and Exertion Activity 127
IV. Summary 132
References 133

Exercise Performance of Birds

DAVID K. SAUNDERS AND M. ROGER FEDDE

I. Introduction 139
II. Energy Demands of Avian Exercise 141
III. Physiological Adaptations for Prolonged Flights and Flights at High Altitude 144
IV. Physiological Adaptations for Diving 172
V. Conditioning and Deconditioning in Birds 176
VI. Summary 179
References 180

Exercise Performance of Mammals: An Allometric Perspective

STAN L. LINDSTEDT AND RICKEY G. THOMAS

I. Introduction 191
II. Body Size and Aerobic Exercise 193
III. Muscle: The Source of Oxygen Demand 196
IV. Oxygen Delivery to Muscles 203
V. Pulmonary Oxygen Uptake 205
VI. Kinetics of Oxygen Transfer 206
VII. Body Size and Aerobic Performance: Emergent Properties 210
VIII. Summary 213
References 213

Beyond the Vertebrates: Achieving Maximum Power during Flight in Insects and Hummingbirds

DOMINIC J. WELLS AND CHARLES P. ELLINGTON

I. Introduction 219
II. Metabolic Power Input 220
III. Muscle Power Output 224
IV. Maximal Muscle Power Output 228
V. Summary 229
References 230

INDEX 233

CONTRIBUTORS

Numbers in parentheses indicate the pages on which the authors' contributions begin.

ALBERT F. BENNETT, Department of Ecology and Evolutionary Biology, University of California, Irvine, Irvine, California 92717 (113)

CHARLES P. ELLINGTON, Department of Zoology, University of Cambridge, Cambridge CB2 3EJ, United Kingdom (219)

M. ROGER FEDDE, Department of Anatomy and Physiology, College of Veterinary Medicine, Kansas State University, Manhattan, Kansas 66506 (139)

STAN L. LINDSTEDT, Department of Biological Sciences, Northern Arizona University, Flagstaff, Arizona 86011 (191)

RICHARD L. MARSH, Department of Biology, Northeastern University, Boston, Massachusetts 02115 (51)

DAVID K. SAUNDERS, Department of Anatomy and Physiology, College of Veterinary Medicine, Kansas State University, Manhattan, Kansas 66506 (139)

RICKEY G. THOMAS, Department of Biological Sciences, Northern Arizona University, Flagstaff, Arizona 86011 (191)

PAUL W. WEBB, Department of Biology, and School of Natural Resources and Environment, University of Michigan, Ann Arbor, Michigan 48109 (1)

DOMINIC J. WELLS, Department of Veterinary Basic Sciences, Royal Veterinary College, London NWI OTU, United Kingdom (219)

PREFACE

Comparative physiology is an integrative approach to understanding how animals function. Its roots are firmly anchored in "whole animal" physiology, a perspective that is essential for interpreting and comparing data collected from specific organ systems in sometimes very different animals. The demands placed on an animal due to environment, body size, etc., are inevitably met by the integrated responses of multiple systems, none of which is independent of other systems in the body. Only by maintaining an integrative view can many of these relationships be understood and their roles in the adaptation and evolution of the animal be interpreted.

Comparative physiologists have two fundamental goals when investigating how animals function. One goal is to determine what structural and functional elements of animals are relatively invariant among taxa, thereby identifying the fundamental building blocks of animals that have not been particularly malleable during the course of evolution. An alternative, but complementary, goal is to understand the mechanisms that enable certain species to tolerate extreme environments or to achieve exceptional physiological performance and to contrast these extreme cases with other animals. This approach capitalizes on the evolutionary variation in structure and function that occurs in different organ systems and provides a large signal-to-noise ratio in data that are collected. Both of these goals contribute to our understanding of how animals adapt to demands of the environment and, over time, evolve in response to those demands.

During exercise, many organ systems approach or attain their maximal functional capacities, and structural elements may be stressed to their limits. Maximally functioning organs or systems may interact in unique ways, resulting in unusual physiological phenomena. The comparative approach is useful for understanding the basic elements that enable, as well as limit, exercise performance.

This book and a companion volume (Volumes 38A and 38B) apply both of the fundamental research approaches of comparative physiology to the study of exercise. This volume explores extremes of adaptation that enable some animals to perform athletic feats that are truly remarkable by human standards. These six chapters deal with aspects of exercise performance that are highly specialized within the vertebrate classes. Extremes of adaptation are discussed and the functions of multiple systems are integrated to explain observed whole animal performance. In the companion volume, principles of structure and function that are similar and fundamental to all vertebrate exercise performance are identified and characterized.

Contributors to this book are authorities in their fields and provide broad perspectives on questions of interest and significance in the study of exercise performance. I am indebted to each of them for the efforts they have made to identify the key elements in their area that are pertinent for comparing exercise performances of vertebrates.

For many of us, our interest in comparative physiology was either initiated or heightened by reading about phenomena such as how kangaroo rats and camels tolerate desert heat and dryness, how birds breathe, or why small animals metabolize faster than large ones. We have also been deeply touched and inspired by the curiosity, creativity, and ingenuity of one man who asked and answered those questions of how animals function. This book is dedicated to that man, Knut Schmidt-Nielsen.

James H. Jones

Exercise Performance of Fish

PAUL W. WEBB

Department of Biology and School of Natural Resources and Environment, University of Michigan, Ann Arbor, Michigan 48109

I. Introduction
II. What Is Performance?
  A. Mechanics
  B. Energetic Performance
III. Locomotor Range Fractionation
  A. Locomotor Range Fractionation
  B. Locomotor System Components
  C. The Multigear Propulsion System
IV. Classification of Performance Levels
  A. Endurance Curves
  B. Factors Determining the Shape of the Endurance Relationship
  C. The Complete Endurance Relationship
  D. Evolution of Gaits and Performance Range
  E. Species Variation in Gaits and Performance Range
V. Limits to Performance
  A. Speed
  B. Acceleration
VI. Conclusions
  References

I. Introduction

The locomotor performance of fishes has fascinated humans for centuries, and the cause for such wonder is easily appreciated by any who have seen salmon leaping waterfalls or manta rays leaping from the sea. Interest in swimming performance in this century was spurred by Sir James Gray, who was unable to reconcile the energy balance for a dolphin using best estimates of muscle power output, swimming speed, and drag (Gray, 1936). Following this observation, dubbed "Gray's

Paradox," numerous studies sought to determine maximum speed and explore other aspects of swimming physiology that might explain the superior performance of fish compared to human-engineered bodies. A careful analysis by Bainbridge (1961) using more up-to-date information essentially resolved the paradox, but not the interest in the magnitude and determinants of fish performance (see Lighthill, 1975; Webb, 1975a; Wu *et al.*, 1975; Pedley, 1977; Blake, 1983a).

Fish originated at least as early as the Silurian, and with their long evolutionary history they have evolved diverse locomotor mechanisms. Among the approximately 20,000 extant species, every possible appendage is used to swim, as well as for gliding in air (perhaps also adding some energy to extend a glide), burrowing, crawling over aquatic and terrestrial surfaces, and even climbing trees (Lagler *et al.*, 1977).

All locomotor behaviors are bounded by physical laws, and the physical properties of the medium determine, in large part, the unique aspects of swimming compared to terrestrial locomotion (Daniel and Webb, 1987). Of particular importance is the density ratio, the ratio of the density of an animal to that of the surrounding fluid. In water, the density ratio is small, close to 1, compared to about 1000 in air. As a result, weight in water is small, permitting the evolution of diverse propulsors and muscle systems for thrust rather than weight support (Webb, 1988a). In addition, it is easy to generate forces that are large compared to the body mass, permitting high rates of acceleration (Daniel and Webb, 1987; Webb, 1988a). On the other hand, it is difficult to hold station on the bottom in a current.

Physical properties of water affect swimming performance in other ways. Because of the high thermal capacity of water, it is hard for a gill breather to achieve endothermy and the associated high rates of energy transduction (Bennett, 1980, 1991). Oxygen resources fueling sustainable locomotion are poor because of the low solubility of oxygen in water, a problem exacerbated by temperature and dissolved materials. Energy costs can be elevated by toxicants that readily dissolve in water. Thus, the scope of possible biomechanical and physiological factors affecting performance is large.

In spite of many years of intense study, knowledge of factors determining and limiting performance of fish is remarkably fragmentary and incomplete. A major reason is a legacy from Gray's Paradox—a focus on studies of maximum performance. This is achieved when swimming with the body and caudal fin, termed *BCF propulsion*. Swimming performance using median and paired fins (*MPF propulsion*) has received little attention, in spite of the ubiquitous use of such systems. Integration of various MPF and BCF systems by fish has not been considered.

This paper summarizes the state of the art of enquiry into exercise performance for fish. It does not focus on data. Instead I consider how fish performance is studied and the difficulties encountered, and suggest possible approaches to ameliorate problems and provide a more comprehensive understanding of swimming performance patterns. The first part of this paper considers the parameters needed to describe swimming performance. The second part discusses the problem of efficiently powering swimming over a very large performance range. I use Alexander's (1989) application of gaits to swimming, encompassing all means of swimming, as well as ecological and evolutionary patterns. The last part considers levels of performance using various gaits, focusing on problems of measurement and interpretation of performance limits.

II. What Is Performance?

A. Mechanics

1. *Motor Performance*

"Swimming" is an imprecise term that applies to many complex behaviors. In every situation, however, the translation of the body can be described in terms of a state variable, speed (u), and control variables, linear acceleration (du/dt) and centripetal (turning) acceleration (u^2/r), where r is the turning radius. Therefore, motor performance can be quantified in terms of u, du/dt, and r.

A system of equations in three dimensions is required to fully describe swimming. In practice, limitations associated with lateral bending of the body and behavioral choices allow swimming to be reasonably described in two-dimensional planes; experimental studies invariably confine fish to such a plane.

2. *Performance and the Force Balance*

Any motion is resisted by the water, and the inertia of the body increases resistance to accelerations (Daniel and Webb, 1987). A resistance component can be linked to each state and control variable. Drag, R_{drag}, depends on velocity:

$$R_{\mathrm{drag}} = 0.5\ \rho\ S\ u^2\ C_d, \tag{1}$$

where ρ = density of water; S = reference area, historically wetted surface area; C_d = empirically determined drag coefficient; swimming at constant average speed using the body and caudal fin increases C_d by a factor of 3 to 5 over that of a rigid reference, usually a flat plate of equal length and area oriented parallel to the flow (Alexander, 1967; Lighthill, 1975; Webb, 1975a; Blake, 1983a).

The resistance components associated with the control variables are linear acceleration resistance, R_{lin}, and centripetal acceleration resistance, R_{cent},

$$R_{\text{lin}} = 0.5\, M\, (1 + \alpha)\, du/dt \tag{2}$$

$$R_{\text{cent}} = 0.5\, M\, (1 + \beta)\, u^2/r, \tag{3}$$

where α = empirically determined added mass coefficient for linear acceleration; it is equivalent to the mass of water that would need to be accelerated with the body to give the observed force at the observed acceleration rate, and a value of about 0.2 probably applies to fish (Webb, 1982a); β = empirically determined added mass coefficient for centripetal acceleration, analogous to α, and also probably about 0.2 (Webb, 1982a); M = body mass.

The vector sum of these resistance components is the *net resistance*, R, given by

$$R = \sqrt{(R_{\text{drag}} + R_{\text{lin}})^2 + R_{\text{cent}}^2}\,. \tag{4}$$

The net resistance is equal at every instant to a thrust force, T, generated by the propulsors. The relationship between thrust and resistance is the force balance, a fundamental relationship for the analysis of locomotion problems (Daniel and Webb, 1987; Webb, 1988a), where

$$T = R = \sqrt{(a_1 u^2 + a_2 du/dt)^2 + (a_3 u^2/r)^2} \tag{5}$$

and a_1, a_2, and a_3 are fitted constants (Webb, 1991).

Maximum thrust is set by the physiological properties of the muscle systems powering the propulsors and the mechanical properties of those propulsors. Maximum values for the motor performance parameters can be calculated by equating physiologically maximum forces and rates of working to resistance, or the power required to overcome resistance [Eqs. (1)–(5)]. Such calculations have been performed frequently (e.g., Bainbridge, 1961; Webb, 1975a; Blake, 1983a; Webb *et al.*, 1984; Somero and Childress, 1980; Goolish, 1991), but always with the implicit assumption that one or more of the resistance components is zero. For example, Bainbridge (1961) calculated maximum speeds by equating muscle power to drag power and solving for u. All such analyses make the implicit assumption that $du/dt = 0$ and $r = \infty$. Theoretical maximum speeds have been calculated for fish accelerating from rest at various mean rates assuming a limited work capacity for muscle (Webb, 1975a). This analysis implicitly assumes $r = \infty$.

For real fish motions, u, du/dt, and r can co-vary in a wide range of maneuvers [Eq. (5)]. Therefore, at the physiological limit of maximum force or power output, the magnitude of a given motor performance parameter will depend on the magnitudes of the others. As a result, maximum speed attainable in maneuvers should be lower than that in rectilinear sprinting. Future reports of maximum performance should specify the values of all three motor performance parameters in as many planes as are relevant to the situation.

Studies of submaximal swimming activities have usually reported only mean speed, even though fish usually accelerate and decelerate during routine swimming (McCutcheon, 1971; Nursall, 1958; Webb, 1991). Weihs (1981) considered the consequences of swimming in a circular path on the force balance in order to correct speeds for energy expended in overcoming centripetal resistance. He assumed resistance vectors were additive, which, on the basis of subsequent studies, may not be correct (Webb, 1991). Reports on submaximal swimming activities should provide more complete kinematic descriptions of motor performance variables.

Thrust and drag involve both inertial and viscous forces. The relative importance of these is roughly indicated by Reynolds number,

$$\mathrm{Re} = (L\ u)/\nu, \tag{6}$$

where L = characteristic length, typically total length for fish; ν = kinematic viscosity of water.

Thus, Eq. (1) assumes Re is high enough that inertial forces dominate the force balance. Many larval fish swim routinely at Re < 30 (Fuiman and Webb, 1988), where for practical purposes viscous forces dominate (Weihs, 1980; Webb and Weihs, 1986). In this case, resistance components in Eqs. (4) and (5) may be reduced to R_{drag}, but the appropriate form of the drag equation and magnitude of the drag coefficient for fish swimming at low Re are not clear (Vlymen, 1974).

B. Energetic Performance

Energy is required to move the body. Rates of energy expenditure required to power different speeds and acceleration rates provide another common measure of *energetic performance*. Energetic performance is a measure of the physiological capabilities of a fish for swimming, also determining limits of motor performance as noted earlier.

Energetic performance can also be examined in terms of energy-minimizing behaviors. These include optimal cruising and foraging speeds, burst-and-coast swimming, burst-and-glide swimming, and selective tidal stream transport (Weihs and Webb, 1983). Also included are mechanisms to avoid swimming, especially when holding position in currents at zero ground speed.

III. Locomotor Range Fractionation

A. Locomotor Range Fractionation

The mechanical power requirements for swimming increase with performance level. The increase is especially rapid for swimming speed, because power requirements increase with u^3 [Eq. (1)]. For example, a salmonid 25 cm in total length might be capable of BCF swimming at speeds from 10 to 250 cm s^{-1} (Brett, 1965; Wardle, 1975). To do this, the myotomal muscle would have to generate forces spanning a 625-fold range ($250^2/10^2$) and power spanning a 15,625 ($250^3/10^3$) range. Muscle and propulsor efficiencies increase roughly linearly with speed (Weihs, 1973a, 1974), certainly with an exponent of u very much smaller than 3. Thus, a major problem for fish is providing thrust and power with reasonable efficiency over a large motor performance range. This problem is much greater for fish than for terrestrial pedestrians, for which power requirements increase linearly with speed, and for flyers, for which the increase is approximately U-shaped.

Pedestrians and flyers meet their locomotor power requirements with a single propulsor. An important design feature for these propulsors is the support of body weight. In contrast, the low density ratio for fish greatly reduces the problem of weight support (Daniel and Webb, 1987). As a result, fish swim using all possible axial and appendicular appendages, so that most have multiple propulsors or, following Lighthill (1975), multiple swimming modes.

Terrestrial animals power legs and wings with up to three types of muscle fibers. These have different functional properties matched to locomotor requirements to provide efficient and effective use of muscle force and power (Rome *et al.*, 1988; Alexander, 1991).

Fish also have up to three types of muscle fibers driving their propulsors (Rome *et al.*, 1988), however, discrete muscle systems are connected to each propulsor. In addition, large masses of muscle drive some propulsors. For example, the myotomal muscle may be 60% of the body mass (Bainbridge, 1962; Webb, 1975a). Therefore, not only do fish have many propulsors, but muscles can be matched to mechanical needs of each (Archer and Johnston, 1991), providing multiple propulsor systems.

Terrestrial pedestrians also use a variety of gaits to optimize performance over a range of speeds (Alexander, 1989, 1991). Alexander (1989) defines a gait as "a pattern of locomotion characteristic of a limited range of speeds described by quantities of which one or more change discontinuously at transitions to other gaits." Alexander formulated this definition to recognize the common occurrence of locomotor transitions among animals, including flyers and swimmers. Following Alexander's definition, gaits in fish may be associated with use of (a) different swimming modes, (b) different muscle fibers driving each mode, or (c) different locomotor behaviors, for example, burst-and-coast swimming.

The numerous gaits of fishes compared with the small number used by terrestrial animals provide the solution to effective and usually efficient swimming over a large performance range. Borrowing a well-known concept from sensory physiology (Eckert, 1988), gaits represent a multigear system for locomotor performance range fractionation, where individual gaits work over a part of the range, and a series of gaits are recruited to cover the entire range (Webb, 1994). Before considering how gaits are used for performance range fractionation, the nature of propulsor system components must be considered.

B. Locomotor System Components

1. *Muscle Types*

The three primary types of swimming muscle differ in stress, power output, and fatigue resistance (Rome *et al.*, 1988). Fast phasic (white) muscle uses phosphagens and glycolysis as energy sources, developing high stresses and high power output but quickly fatiguing. Low mitochondrial density and poor vascularization of this muscle also result in slow recovery from fatigue. In contrast, slow tonic (red) muscle develops low stress. Red muscle is aerobic, with high mitochondrial density, capillary density, and myoglobin concentration, and is nonfatiguing. Fish also have small amounts of fast, phasic, aerobic (pink) muscle, presumably intermediate between red and white muscle in endurance, stress, and power output (Johnston, 1981, 1991). As with other vertebrates, the various muscle types are recruited sequentially from red through pink to white as locomotor power requirements increase (Johnston *et al.*, 1977; Goldspink, 1981; Johnston, 1991). Each muscle recruitment step represents a discontinuity, and hence a gait transition.

2. Propulsors

The muscles drive propulsors that transfer momentum from the fish to the water, thereby propelling the body (Daniel and Webb, 1987; Webb, 1988a). Structural and functional diversity among fish propulsors is large, and numerous classifications have attempted to find common features (Breder, 1926; Lindsey, 1978; Braun and Reif, 1985; Webb and Blake, 1985). The early classification by Breder (1926) was largely morphological, only secondarily considering propulsor movements. Breder named various propulsor categories with prefixes derived from the family or a group that emphasized the use of a particular fin or locomotor pattern and the suffix "form," for example, anguilliform, after the family Anguillidae, meaning eellike (Fig. 1). This precedent is still used, but recent classifications seek common patterns in mechanical principles and kinematics (Lindsey, 1978; Braun and Reif, 1985; Webb and Blake, 1985; Daniel and Webb, 1987; Webb, 1988a).

Most classifications distinguish between undulatory and oscillatory propulsion (Lindsey, 1978; Braun and Reif, 1985; Webb and Blake, 1985). There have been numerous studies of BCF modes, so Breder's (1926) classification has been refined and modes are now understood in terms of kinematics and the resulting major forces contributing to thrust. Thus, the undulatory anguilliform mode uses resistive and reactive forces (Lighthill, 1975). Reactive (acceleration reaction) forces are most important in the undulatory subcarangiform and oscillatory ostraciiform modes (Blake, 1977, 1981; Daniel, 1984). Reactive and lift forces are important for carangiform swimmers, whereas lift forces are most important in the oscillatory thunniform mode (Lighthill, 1975). Efficiency is believed to increase with swimming mode as

$$\text{ostraciform} < \text{anguilliform} < \text{subcarangiform} < \text{carangiform} < \text{thunniform}.$$

The gradation from undulatory to oscillatory mechanisms is mirrored among MPF propulsors. Reactive forces are probably most important in undulation of long-based median fins (amiiform, gymnotiform, balistiform, and rajiform modes) (Lighthill and Blake, 1989; Lighthill, 1990a,b,c). Short-based truncated fins probably use resistive and reactive forces in both undulation (diodontiform mode) and oscillation (rowing; labriform mode), whereas short-based elongate fins probably act as hydrofoils (labriform and tetraodontiform modes).

BCF modes have clear hydromechanical and functional meaning. A

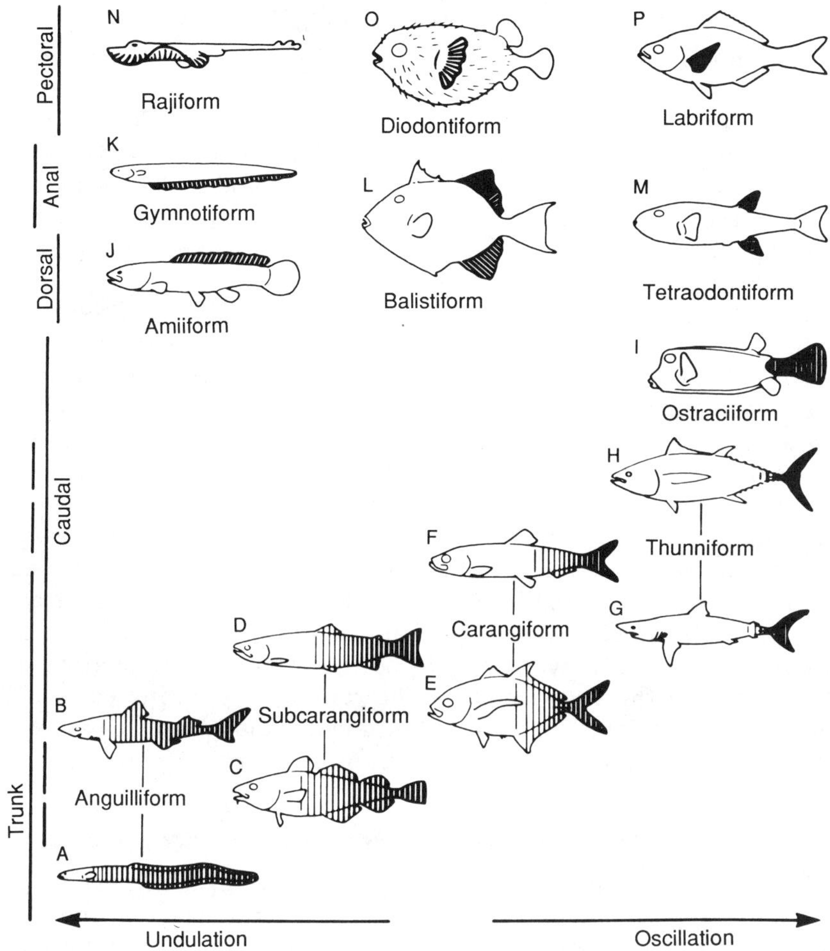

FIG. 1. A summary of some major features of classifications of fish propulsors, illustrating the historical approach based on morphology (left) and an approach based on undulation and oscillation. From Lindsey (1978).

similar understanding and replacement of the Breder classification is necessary for MPF swimming. However, here the situation is complicated because of the flexibility of fins; for example, the labriform mode encompasses drag-based and lift-based mechanisms (Webb and Blake, 1985), pectoral fins of different rays may undulate or oscillate (Daniel, 1988), and the pectoral fins of centrarchids may undulate or oscillate in different activities. Eventually all MPF modes should be described in terms of mechanical properties, examining the interplay of key kinematic variables, as Daniel (1988) did for rays.

3. *Behavior Patterns*

Swimming behavior may be "steady" or unsteady. In the former case, speed variation about an average speed is small and due to variations in thrust within a tail-beat cycle (DuBois *et al.*, 1976; Freadman, 1981; Videler, 1981; Weihs and Webb, 1983; Webb, 1988b). Speed is variable in unsteady swimming, so that linear and centripetal accelerations become essential to describe the motion. Unsteady motion has been used to define two different BCF locomotor behavior patterns: intermittent swimming and fast-starts.

Intermittent swimming consists of a few tail-beats accelerating a fish, followed by an unpowered coast, or a glide by negatively buoyant fish (Weihs, 1973a, 1974; Videler and Weihs, 1982). Because swimming resistance is 3 to 5 times greater than gliding resistance, intermittent swimming can reduce energy requirements by over 50% compared to those during continuous steady swimming at the same average speed (Weihs, 1974). Therefore, this behavior is seen when endurance is important.

Fast-starts are maximum accelerations. They have only been examined for fish accelerating from rest, but kinematically analogous accelerations occur during steady swimming. Fast-starts are mechanically distinct from steady swimming, as body movements are nonrepeating and forces on the body and fins anterior to the trailing edge of the caudal fin contribute to both instantaneous and mean thrust (Weihs 1972, 1973b; Webb, 1982b, 1984a; Weihs and Webb, 1983).

C. The Multigear Propulsion System

Individual fish use the locomotor system components in various gaits, recruited in an orderly sequence as performance level increases, especially for increases in speed (Webb, 1994). At low performance levels, small amounts of red muscle drive MPF propulsors (Fig. 2). Little muscle is needed because speeds and acceleration rates are small. In addition, the body is held rigid, avoiding the drag increment associated with body undulation. Thrust enhancement (Lighthill, 1990b) may also reduce red muscle needs.

Three MPF gaits are possible using red, pink, or white muscle fibers, but red fibers usually dominate (Hochachka and Somero, 1984; Davison and MacDonald, 1985; Archer and Johnston, 1991). In labriform swimmers, the white fibers may be important not for propulsion but to quickly adduct the fins to reduce drag in BCF swimming (Harrison *et al.*, 1987; Johnston, 1991). As a result, gait transitions during MPF propulsion are unlikely to depend on muscle recruitment.

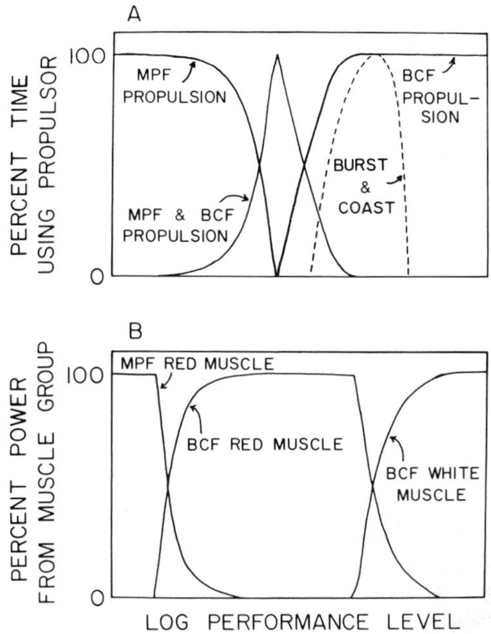

FIG. 2. Diagrammatic illustration of fractionation of the speed and acceleration performance range by (A) propulsors and (B) muscle type. Each different combination of propulsor and muscle type defines a gait. Note that performance is shown on a logarithmic scale to expand the lower performance range where gaits are more numerous.

Several MPF gaits are possible that are associated with differences in kinematics, for example, shifts from diodontiform to labriform modes, but data are too sparse to evaluate this possibility.

As speed increases, MPF propulsion is supplemented and then succeeded by BCF propulsion (Figs. 2 and 8). Within BCF propulsion, additional discontinuities are associated with the various muscle types and with steady, unsteady, and intermittent swimming.

At lower BCF swimming speeds, the propulsor is driven by myotomal red muscle (Fig. 2). Because speeds are low, relatively little red muscle is required, and it constitutes 5 to 7% of body mass (Bone, 1966; Johnston, 1981; McLaughlin and Kramer, 1991). As speed increases further, white muscle first supplements, and then succeeds, red muscle. This large mass of muscle is essential to meet the high power requirements of sprinting and accelerating at high levels of performance.

With the recruitment of anaerobic white muscle, fish begin to swim at speeds at which they eventually fatigue. Intermittent swimming appears to be prevalent in the lower range of fatigue speeds (Videler and Weihs, 1982; Alexander, 1989), but in theory might be expected at all speeds (Alexander, 1989). In practice, intermittent swimming would not be effective for activities calling for maximum speed (e.g., to chase elusive prey, to avoid capture by chasing predators, or to negotiate short regions of high flow) because the average speed, by definition, must be lower than the maximum. Intermittent swimming does occur as part of the routine swimming repertoire of schooling species (Nursall, 1973) and in saltatory foraging (O'Brien *et al.*, 1986), but no analysis (e.g., Videler and Weihs, 1982) has been done to determine if these patterns meet criteria for energy economy. Intermittent swimming has also been described in carp at very low speeds to offset low muscle efficiency (Rome *et al.*, 1990). However, in general, intermittent swimming does not appear to be common at low swimming speeds, perhaps because intermittent swimming can be a predator-releasing stimulus (Harper and Blake, 1991; P. W. Webb, unpublished results).

Energy requirements for centripetal acceleration increase with u^2/r, and as r is independent of u (Howland, 1974; Webb, 1983), forces and power required for turning would follow those for speed. Therefore, turning maneuvers at different speeds should show similar gait patterns as speed (Fig. 2).

Energy requirements for linear acceleration increase linearly with du/dt. However, high acceleration rates require large-amplitude nonrepeated movements characteristic of a fast-start, and hence a different gait. In addition, these high accelerations involve a turn. As a result, the real energy costs for linear acceleration may also span a range similar to that for speed. In general, therefore, similar range fractionation by gait is expected for accelerations and speed, except that high acceleration rates require the unique fast-start gait.

The application of Alexander's (1989) ideas on gaits to the full range of swimming performance requires reconsideration of traditional approaches to swimming performance (Webb, 1994). The historical focus on BCF propulsion ignores the great diversity of propulsors among species, as well as within a species, for efficient swimming over a wide performance range.

IV. Classification of Performance Levels

A. Endurance Curves

Studies of swimming usually define performance levels following recommendations by Hoar and Randall (1978; Table I). These were

TABLE I

THE TRADITIONAL CLASSIFICATION OF PERFORMANCE LEVELS FOR FISH SWIMMING, BASED LARGELY ON RECOMMENDATIONS BY HOAR AND RANDALL (1978)

Performance level	Definition	Source
Sustained swimming	Steady speeds maintained indefinitely, or for >200 min; nonfatiguing, aerobic activity	Hoar and Randall (1978)
Prolonged swimming	Steady speeds between sustained speeds and sprints, lasting between 5–10 s and 200 min; performance at this level ultimately results in fatigue; aerobic or anaerobic activity; performance is measured as the critical swimming speed	Brett (1964), Hoar and Randall (1978), Farlinger and Beamish (1977)
Burst swimming		
Sprints	Steady swimming at speeds maintained <5–10 s; anaerobic activity	Hoar and Randall (1978)
Fast-starts	Maximum accelerations, usually from rest, lasting <1 s; anaerobic activity	This paper
Routine swimming	Volitional movements at sustained, prolonged, or burst performance levels; aerobic or anaerobic activity, with or without fatigue	Hoar and Randall (1978), Webb (1991)

based on data for strong BCF swimmers (e.g., salmonids) at steady speeds typically maintained for longer than a second, following breaks in the empirical relationship between speed and the time to fatigue at a given speed (Fig. 3).

This classification implicitly assumes a causative role for the properties of the metabolic pathways in the muscles. Energy available from aerobic metabolism is unlimited for practical purposes, but the rate of energy release is relatively low. Therefore, there is a threshold at low speeds below which endurance is infinite and independent of speed. This defines sustained or cruising speeds (Table I). In contrast, capacity limits for anaerobic metabolism contribute to decreasing endurance at high speeds, defined as sprints. Between sustained and sprint speeds is a range of speeds that can be sustained for considerable periods, but at which fatigue eventually occurs. These are prolonged speeds (Table I).

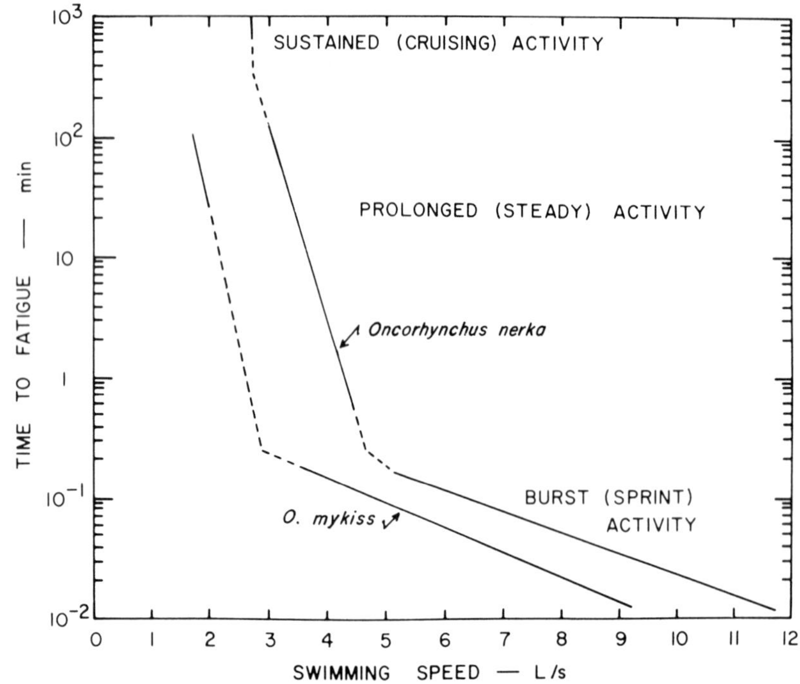

FIG. 3. Examples of the endurance–speed relationship measured for salmonid fishes. Modified from Webb (1975a).

B. Factors Determining the Shape of the Endurance Relationship

Several factors will affect the shape of the endurance–speed curve. Eventually, endurance must depend on the rate of energy use and the size of the energy pool available from anaerobic metabolism (Bennett, 1980, 1991; Somero and Childress, 1980; Hochachka and Somero, 1984). If the energy capacity from hydrolysis of phosphagens or from glycolysis is E, then based on Eq. (1),

$$E = 0.5\ \rho\ S\ u^3\ C_d\, t, \tag{7}$$

where t = endurance in seconds. From Eq. (7), endurance would be related to speed by the negative power function

$$t \propto u^{-0.33}. \tag{8}$$

In practice, although endurance is inversely related to speed, the data are best described by an exponential function (Fig. 3; Videler and Wardle, 1991)

$$t = ae^{bu}. \tag{9}$$

The observed endurance–speed relationship (Fig. 3) has three sections, the number expected if phosphagens, glycolysis, and then aerobic pathways fuel swimming at different speeds. However, the observed sections do not correlate with those expected. Very high sprint speeds supported by hydrolysis of phosphagens may not have been measured, but if they have, there is no break in the endurance–speed relationship for sprints. Instead, there is a break in the relationship between sprints and sustainable speeds. Therefore, metabolic capacity alone is insufficient to explain the relationship between endurance and speed.

Additional factors of importance are:

1. Overlapping use of metabolic pathways in muscles (Bone, 1966; Johnston *et al.*, 1977; Johnston, 1981);
2. Use of variable numbers of motor units;
3. Use of various gaits (Alexander, 1989, 1991);
4. Varying contributions of resistance components (Daniel and Webb, 1987; Webb, 1988a);
5. Muscle working on different parts of the muscle stress-strain curve (see paper by Rome in Volume 38A).

These factors could generate numerous endurance–speed relationships. Indeed, for prolonged swimming, the exponent b in Eq. (9) varies from -0.14 to -2.1 (Videler and Wardle, 1991). Fewer factors are likely to influence endurance at maximum sprint speeds, and there is less variability in b, which varies from -0.13 to -0.34 (Videler and Wardle, 1991).

Experimental studies to evaluate factors that determine endurance in different species are lacking. Comparative studies are needed that pay greater attention to swimming behavior, as well as physiological changes that occur in various carefully measured speed ranges for fish with differing compositions of myotomal muscle (see Goolish, 1991). Experimental protocols are needed to examine sprint performance in a controlled fashion over a range of speeds, as well as in transition from fast-starts to sprints. A possible role of control systems determining endurance should also be considered, but as fish can be exercised to a degree of exhaustion that causes death (Wood *et al.*, 1983; Wood, 1991), these are probably less important.

C. The Complete Endurance Relationship

The foregoing discussion on performance range fractionation by gaits and the nature of the metabolic pathways in different muscle systems can be combined into a generic, multisectional endurance–performance relationship that spans the full range of swimming activities of fish with sections delimited by gait transitions (Fig. 4). Speed may be zero relative to the water or relative to the ground, and these are added to the performance range.

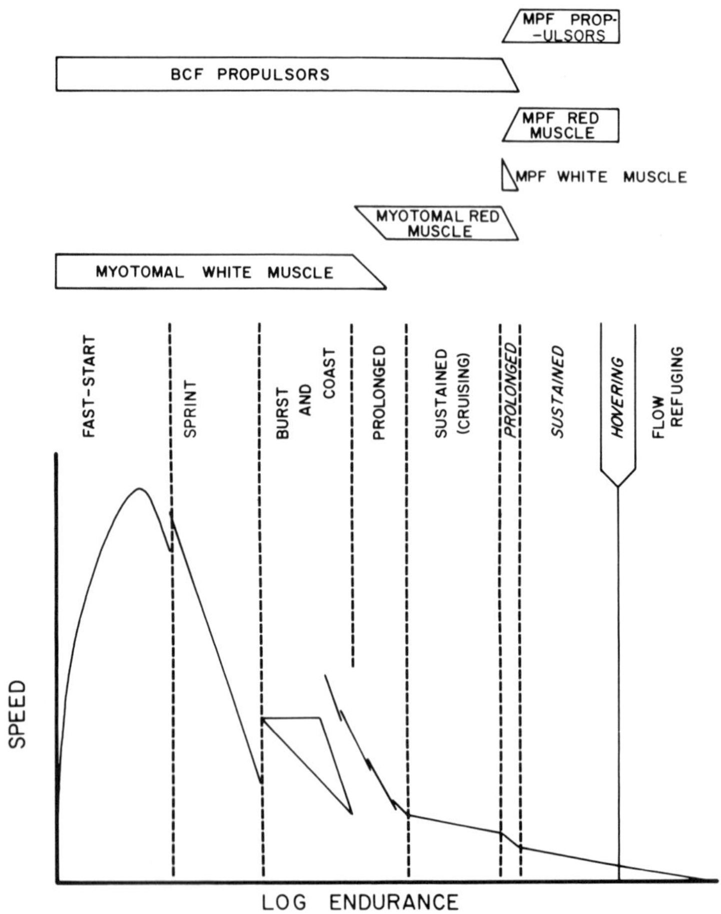

FIG. 4. A hypothetical generic endurance–speed relationship for swimming fishes. The relationship is composed of a number of sections, one for each gait. Note that the endurance scale is logarithmic. As with pedestrians, gaits can overlap. Most overlap seems probable between body and caudal fin (BCF) prolonged and intermittent swimming. A number of endurance relationships with decreasing slope are illustrated for

At low swimming speeds, either forward or backward (Blake, 1980, 1983b), the generic fish is propelled by median and paired fin propulsors driven by red muscle, as described earlier. This is succeeded by the various BCF gaits, and eventually fast-starts as performance level increases.

I suggest that the shape of the endurance–performance relationship within gaits is determined by a limiting capacity of anaerobic pathways and the use of variable numbers of motor units (Bone, 1966, 1977; Eaton *et al.*, 1988; Foreman and Eaton, 1990). I further suggest that very high sprints are limited by the time, and perhaps energy, required to accelerate. This results in speed increasing during, and for some time after, a fast-start to reach a maximum. Thereafter, sprint speed declines according to the time taken to deplete anaerobic energy reserves.

D. Evolution of Gaits and Performance Range

Endurance–performance capabilities and performance range are expected to vary among species depending on the degree of development of the different musculo-propulsor systems underlying various gaits (Webb, 1994). Thus, some species can use all gaits and perform over a wide range of speeds and acceleration rates. Others lack the morphology to swim in some gaits, and their performance range is diminished accordingly.

The number of swimming gaits expressed by fish has increased through evolutionary time. The earliest Paleozoic agnathan fishes were typically elongate and armored, and paired fins were stiff or absent (Moy-Thomas and Miles, 1971). Small body depth and armor undoubtedly reduced the acceleration performance of these fish (Webb and Smith, 1980; Webb *et al.*, 1992). The paired fins probably acted as hydrofoils to control position in the water column, as is the case with modern negatively buoyant fishes (Alexander, 1990), but would not have been very versatile. Therefore, the early fishes were probably restricted to a narrow performance range of steady and intermittent BCF gaits using red, white, and pink muscle.

The early agnathans were succeeded by the Silurian and Devonian anaspids, cephalaspids, and some heterostracans with deeper bodies

BCF prolonged swimming to suggest a shift in dominance from aerobic to anaerobic metabolic pathways, with recruitment of an increasing number of white muscle motor units. The shape of the endurance relationship for intermittent swimming could take virtually any form, depending on the magnitudes of acceleration and deceleration phases and the proportion of time for which the behavior is used.

and caudal fins, and reduced armor. Webb and Smith (1980) have suggested that these traits were important in the evolution of acceleration capability, and the fast-start gait was probably introduced, or at least greatly amplified, at this time.

MPF gaits require flexible fins. A highly flexible rayed fin structure is an apomorphy for actinopterygian fishes (Moy-Thomas and Miles, 1971). As a result, MPF gaits probably arose with these fishes. Pectoral fins were originally located anteriorly and ventrally on the body, whereas the pelvic fins were posterior and ventral (Rosen, 1982). A more versatile arrangement with laterally placed pectoral fins and anterior ventrolateral pectoral fins evolved in paracanthopterygian and acanthopterygian (spiny-rayed) fishes (Rosen, 1982). The location of these fins in conjunction with the dorsal and anal fins distributes MPF propulsors around the center of mass. This arrangement probably provided for the largest range of MPF gaits. When coupled with BCF gaits, a large number of MPF gaits permitted the largest performance range and undoubtedly has contributed to the successful radiation of the acanthopterygian fishes.

E. Species Variation in Gaits and Performance Range

The occurrence and importance of different gaits vary among fish species, just as they do for pedestrians. For fishes, gaits and performance range vary according to habitat and life-style, including the effect of size (Webb, 1994).

Size effects for large fish originate from the scaling relationship between muscle stress and inertial resistance, and lead to gait compression. Muscle stress and power increase as $M_b^{0.67}$, where M_b is body mass, whereas inertial resistance to acceleration increases as $M_b^{1.0}$. As a result, acceleration performance should decrease in large fish (Daniel and Webb, 1987; Webb and de Buffrénil, 1990). Although the size effect may be offset by physiological adaptations facilitating high speed, for example, endothermy (Bennett, 1980), the fast-start gait is probably less important for the largest fish (Webb and de Buffrénil, 1990). These fish also swim continuously, using hydrodynamic lift to support their body weight in the water column (Magnuson, 1978; Alexander, 1990). The median and paired fins of these fish are stiff hydrofoils (Magnuson, 1978), reducing the utility of MPF gaits.

Larval fish also have reduced gait ranges because their routine swimming is at low Re, where viscous forces dominate. This may affect development patterns of median and paired fins, as well as the magnitudes of inertial forces for acceleration during ontogeny (Webb and

Carolla, 1981; Webb and Weihs, 1986). Therefore, larvae primarily use BCF gaits.

The relationships between the range of gaits expressed by a fish and its habits and habitat are quite variable, but essentially follow the patterns discussed elsewhere for swimming modes (e.g., Alexander, 1967; Sharp and Pirages, 1978; Webb, 1982b, 1984a; Weihs and Webb, 1983). Fast-start performance may be generally reduced in fish with defense mechanisms that minimize the importance of escape by maneuver, for example, burrowing eellike fishes with physical refuges (Eaton *et al.*, 1977), large fishes (Webb and de Buffrénil, 1990), armored fishes (Webb *et al.*, 1992), and probably poisonous fishes. Acceleration gaits may also be reduced in fish in which well-developed inertial suction feeding replaces whole-body acceleration for prey capture, for example, anglerfish (Grobecker and Pietsch, 1979).

Most fish have one or more BCF gaits for sprinting and escape responses. At lower speeds, red-muscle-powered BCF propulsion is common. However, sustainable BCF gaits are absent from fish with little or no myotomal red muscle, for example, esocids and cottids. These fish are typically sit-and-wait predators.

Some of the more derived actinopterygian fishes, such as surfperch (Embiotocidae) and wrasses (Labridae), cruise using their pectoral fins, whereas others, such as ocean sunfish (Molidae) and triggerfish (Balistidae), cruise using the dorsal and anal fin as a pair. Skates and rays (Batoidimorpha) use pectoral fin propulsion at all speeds, and hence lack all BCF gaits (Fig. 1).

High-speed MPF cruising is similar among fishes. The fins of fish that perform this gait, like those of negatively buoyant fish, tend to be elongate, high-aspect-ratio hydrofoils. Their greater stiffness may reduce versatility, especially for low-speed, finely controlled maneuvers. High-speed MPF swimmers do not appear to approach structures as closely as species with less-specialized MPF propulsors (Hobson, 1974), but specific data are lacking. Thus, a lack of versatility may militate against use of MPF propulsion at higher sustainable speeds for most fish.

Because gait variation among species often is associated with habitat and life-style, I assume interspecific differences are adaptive. Therefore, the traditional classification of performance based on fixed endurance times (Table I) should be replaced by an approach that considers endurance in different gaits as traits subject to variation and selection. Such studies should address questions of not only the proximate (physiological, mechanical, and ecological) basis for interspecific variation, but also ultimate causes of selective importance on an evolutionary time scale.

V. Limits to Performance

Swimming performance boundaries have been measured for almost all the gaits proposed here, although there are large data sets only for some BCF gaits. Excellent summaries of data for the latter gaits are provided by Beamish (1978), Harper and Blake (1990), Domenici and Blake (1991), and Videler and Wardle (1991).

A. Speed

1. Zero Ground Speed

Swimming fish are not only exposed to the flow they induce, but in other situations must hold position (zero ground speed) in a flowing environment. Compared to the locomotor problems of terrestrial pedestrians, maintaining zero ground speed in water currents can be problematic. Low density ratio in water results in the fish having a small weight that is easily reduced by lift due to flow over the body. The resultant small force normal to the surface also makes the friction force small relative to the drag force tending to displace the fish downstream (Arnold and Weihs, 1978). However, currents ranging from <10 to >900 cm s^{-1} in streams, rivers, lakes, deep oceans, estuaries, tidal regions, etc., are common features of aquatic habitats. These are often preferred habitats for fish because they are productive and provide for dispersal (Hynes, 1970).

Current-swept habitats are most effectively exploited by fish that avoid the flow and live *adjacent* to the current, entering it only as necessary to catch food, migrate, mate, etc. I define avoidance, or minimization, of swimming in currents as *flow refuging*, and all structures, morphologies, and behaviors used to achieve this end as *flow refuging mechanisms*. Some fish escape currents by using cracks, burrows, or the lee of boulders. The greatest problems associated with flow refuging occur for fish that cannot escape the flow. The overall ability of these fish to avoid flow is measured by the *swim speed*, the maximum speed above which fish must actively swim to hold station (Webb, 1989). This speed is independent of fish size for Re $> 10^4$ (Arnold and Weihs, 1978).

Fish appear to fall into three main flow refuging categories (Table II). Two of these are benthic fishes. Benthic fish with body shapes flattened parallel to the substratum have low drag coefficients but high lift coefficients (Webb, 1989). The extended median fins (Pleuronectiformes) or pectoral fins (Batoidimorpha) delay the need for swim-

TABLE II

FLOW REFUGING PERFORMANCE OF SOME BENTHIC AND BENTHOPELAGIC FISHES, MEASURED AS MAXIMUM CURRENT VELOCITIES AT WHICH FISH CAN MAINTAIN ZERO GROUND SPEED ON SMOOTH SURFACES[a]

Species	Length (cm)	u_{slip} (cm s^{-1})	u_{swim} (cm s^{-1})	Source
Benthic Fishes—low-drag/high-lift forms (pleuronectiform)				
Pleuronectes platessa	18–51	21.7		Arnold and Weihs (1978)
	26	20	58	Webb (1989)
Raja clavata	27*	21	33	P. W. Webb (unpublished
Potamotrygon shroederi	9.3*			observations)
Benthic Fishes—high-drag/low-lift forms (cottiform)				
Percina roanoka	>3.5	30	15–17	Matthews (1985)
Etheostoma flabellare	>3.5	24	15–17	Matthews (1985)
	<2.8	16	15–17	
Myoxocephalus scorpius	24	32	32	Webb (1989)
Otoclinus sp. (Loricariidae)	3.5	26	26	MacDonnell and Blake (1990)
Benthopelagic and Pelagic Fishes				
Gadus morhua	26	6	6	Webb (1989)
Salmo salar	12		81	Arnold *et al.* (1991)
Lepisosteus osseus	27	36	36	Webb *et al.* (1992)
Esox sp. (musky)	19	18	18	

[a] Data are shown for the speed at which they first slip and the speed at which they first swim. Behavioral mechanisms in some species can delay the onset of swimming after fish first slip. Length is total length except for the rays (marked with *), for which length is the length of the body disc, excluding the tail.

ming by generating suction or currents beneath the body to reduce lift (Arnold and Weihs, 1978). Other benthic fish have more fusiform, but somewhat depressed, body shapes that result in high drag coefficients but low lift coefficients (Webb, 1989). These fish often have well-developed pectoral fins that follow the body's contours to reduce drag and increase friction.

The two groups represent different solutions to flow refuging. Pleuronectiform fishes are superior on smoother substratum, but even a small amount of surface texture allows cottiform fish to increase friction and achieve similar swim speeds (Webb, 1989; MacDonnell and Blake, 1990). Pleuronectiform fish are more common where bidirectional tidal flows result in soft bottoms, whereas cottiform fish are more abundant in unidirectional eroding currents where the substratum has greater roughness.

Benthopelagic fish tend to be closer to neutral buoyancy than true benthic fish, with a body shape that gives higher drag coefficients but little lift (Hoerner, 1965, 1975). Such fish have little ability to achieve zero ground speed unless they can capture momentum from the current to generate negative lift (Arnold *et al.*, 1991). Fish using negative lift or high friction devices have the greatest flow refuging performance (highest u_{swim}) and are able to achieve zero ground speeds in currents exceeding prolonged speeds (MacDonnell and Blake, 1990; Arnold *et al.*, 1991).

Performances on substrata with materials and shapes typical of natural habitat are poorly known (Pavlov and Tyurukov, 1988; Pavlov *et al.*, 1982, 1983; Shtaf *et al.*, 1983; Sutterlin and Waddy, 1975), although the ability of plaice to hold station on a rippled surface apparently depends on the relationship of fish length to ripple wavelength (G.P. Arnold, unpublished observations). Flow refuging using boulders, submerged roots, etc., is known to contribute to habitat choices of many stream species (Hynes, 1970), but details of how fish interact with such structures and factors affecting flow refuging performance are poorly understood (Sutterlin and Waddy, 1975).

2. *Zero Water Speed*

Neutrally buoyant fish can hover in the water column. Few, if any, are hydrostatically stable, because the center of buoyancy is usually below the center of mass (Gee, 1983). As a result, continuous fin movements are required to control position, even if there is no weight to support. Performance could be measured as the energy required to control posture (Blake, 1979), which should vary with the propulsor morphology and the anatomy of buoyancy devices determining the magnitudes of rolling and pitching moments (Aleyev, 1977).

3. *MPF Performance*

Few studies have determined the limits of MPF performance. Soft-rayed fish generally appear to shift to BCF gaits at low speeds (Table III). Acanthopterygian fishes with specialized lift-based propulsors achieve the highest levels of sustained and prolonged performance, similar to BCF prolonged speeds. However, as noted earlier, morphological requirements for such high performance appear to reduce versatility for low-speed maneuvering. As a result, most MPF propulsion systems function at low performance levels.

TABLE III

PERFORMANCE LIMITS FOR MPF PROPULSION

Species	Total length (L) (cm)	Maximum speed (using MPF propulsion alone) ($L\ s^{-1}$)	Water temperature (°C)	Source
Lepomis macrochirus	15.6	0.33	15	P. W. Webb, in Alexander (1989)
Lepomis gibbosus	12.7	0.94	20	Brett and Sutherland (1965)
Cymatogaster aggregata	14.3	4.0	15	Webb (1975b)
	9.3	4.6	15–20	Dorn *et al.* (1979)
Hypsurus caryi	13.8	3.0	15–20	Dorn *et al.* (1979)
Hyperprosopon argentum	13.7	4.2	15–20	
Phanerodon furcatus	15.5	3.8	15–20	
Embiotoca jacksoni	14.9	3.6	15–20	
Pagothenica borchgrevinka	23	1.8	−1.8	Montogomery and Mac-Donald (1984)
Notothenia neglecta	7.1–8.1	1.7	2.0	Archer and Johnston (1991)
	26.9–30.5	0.9	2.0	(Fig. 4 and Table I)

4. *Sustained and Prolonged Performance*

Prolonged speeds have often been measured because of the ease of their determination in increasing velocity tests (Brett, 1964). In these tests, swimming speed is increased by a fixed increment, *du*, at regular time intervals, *dt*. For fish with low aerobic scopes (e.g., esocids), time intervals may be as small as one minute. When the objective is to obtain physiological data, time periods are typically on the order of 30 to 60 min. The increasing velocity test continues until a fish is exhausted. A critical speed, u_{crit}, is defined, prefixed by *dt*, and calculated as (Brett, 1964; Beamish, 1978)

$$u_{\text{crit}} = u_{t-1} + u_t\,(t/dt), \tag{10}$$

where u_{t-1} = highest speed at which a fish swims for the entire time interval, *dt*; u_t = speed at which fish finally fatigues; t = time to fatigue at u_t.

Sustained speeds are measured in fatigue tests, in which subsamples swim at various fixed speeds, and times to fatigue are recorded. The time at which half the sample fatigues is then the endurance time (Brett, 1967). The sustained speed is the threshold speed at which endurance becomes infinite. Therefore, determination of maximum sustained speeds is time-consuming, and speeds maintained for 200 min have been used as a convenient substitute. Sustained speeds appear to be 10 to 20% lower than prolonged speeds maintained for about an hour (Brett, 1964; Beamish, 1978).

Values of u_{crit} depend on *dt* and *du* (Fig. 5; Farlinger and Beamish, 1977). For largemouth bass (*Micropterus salmoides*, fork length 9 to 12 cm at 25°C),

$$\begin{aligned}\log u_{\text{crit}} = {} & 1.6 - (0.0043\,dt) - (0.0001\,dt\,du) \\ & + (0.00004\,dt^2) + (0.016\,du) + (0.00061\,du^2).\end{aligned} \tag{11}$$

Thus, highest u_{crit} occurs at *du* of about 1 body length (*L*) when *dt* is large, but larger *du* gives highest speeds when *dt* is small.

Prolonged speeds vary by over an order of magnitude, from approximately 1 to 15 $L\ \text{s}^{-1}$, and similar variation is seen in sustained speeds (Beamish, 1978; Videler and Wardle, 1991). Much of this variation is due to fish size and the environmental conditions at which measurements were made. Videler and Wardle (1991) have summarized the size effects for maximum sustained speed, $u_{\text{sustained}}$, and prolonged speed, $u_{\text{prolonged}}$, to show trends with size:

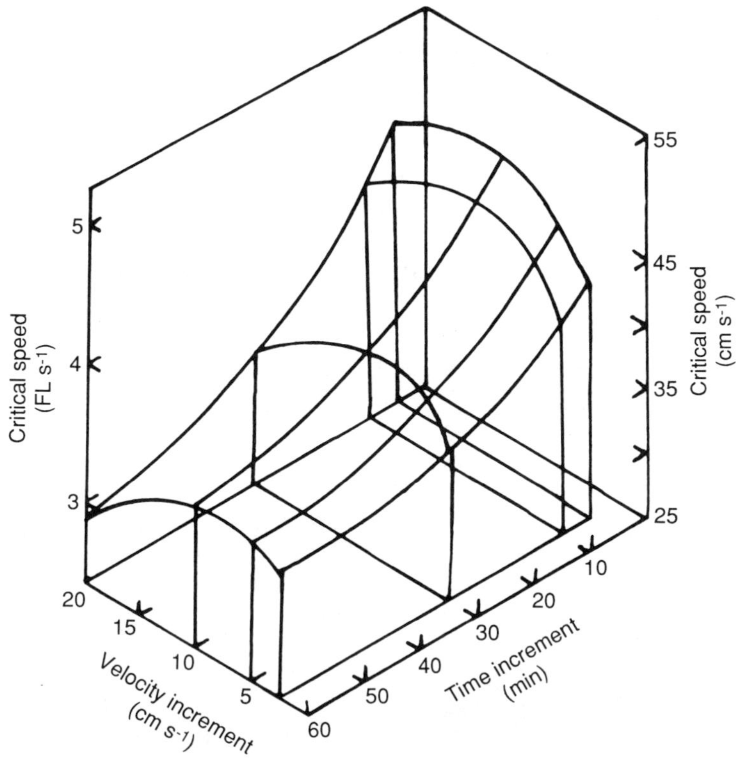

FIG. 5. The effect of speed increment and time interval on the prolonged performance of largemouth bass (*Micropterus salmoides*). From Farlinger and Beamish (1977).

$$u_{\text{sustained}} = 15 + 2.4\,L \tag{12}$$

$$u_{\text{prolonged}} = 61 + 2.4\,L, \tag{13}$$

where speeds are in cm s^{-1} and length is in cm.

The mechanistic basis for the size effect is not fully understood. Limits are probably determined by design characteristics of propulsors, muscles, and metabolite delivery and removal systems, while variation in these systems probably underlies interspecific differences in performance (e.g., Jones, 1971; Jones and Randall, 1978; Randall and Brauner, 1991).

Wardle (1975) has suggested that these scaling relationships are applicable only to fish less than about 1 m in length. He proposed that a jump in power requirements at the boundary-layer transition from laminar to turbulent flow might limit $u_{sustained}$. In that case, maximum $u_{sustained}$ would follow iso-Re lines, and speed would decrease with increasing size (Fig. 6). The iso-Re lines at which limits are reached vary among species depending on the sophistication of body streamlining and temperature, both of which affect the Re at which transition occurs.

Sustained performance, and to a large extent prolonged swimming performance, is regulated by metabolic scope (Beamish, 1978), the difference between standard metabolism (the minimum rate for an un-

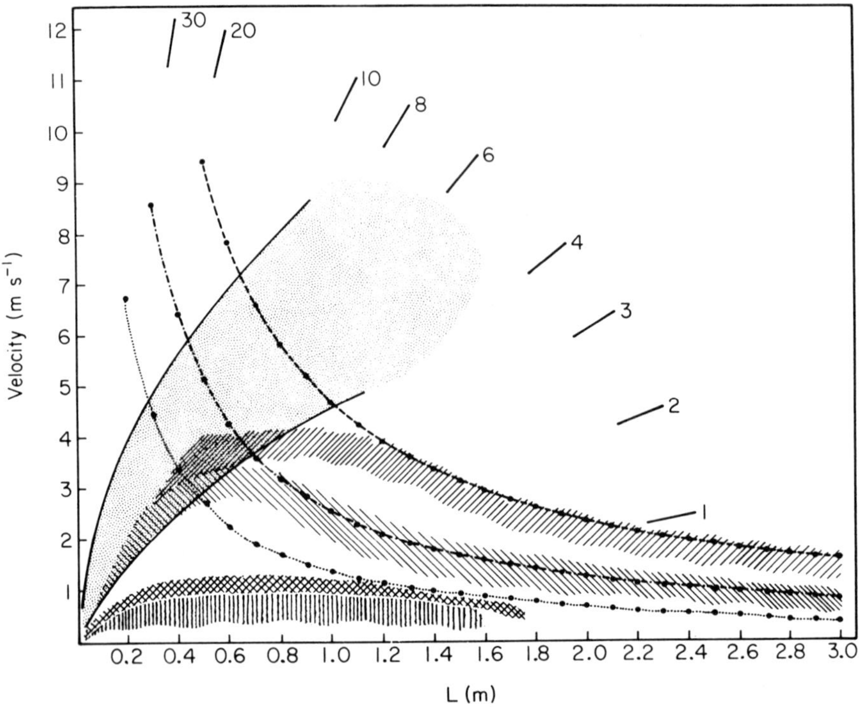

FIG. 6. The relationship between maximum sustained swimming speed (variously dotted lines) and size for a large range of fish size. Sustained speeds of larger fish are postulated to be limited by iso-Re lines, at which boundary-layer flow transition occurs. These iso-Re lines vary among species depending on body form and water temperature: Re = 3×10^6 (5°C), (dashed line); Re = 3×10^6 (30°C), (dash-dotted line); Re = 10^6 (10°C), (dotted line). Speeds are for Salmonidae (⋙), Gadidae (|||), Scombroidei with warm bodies in cold water (///), and Scombroidei with warm bodies in warm water (\\\\\\). Sprint speeds are shown by the stippled area and the third, curved scale from 1 to 30 lengths s^{-1}. These speeds are used to briefly cut through cruising speed size constraints. From Wardle (1977).

stressed, postprandial fish) and active metabolism (the maximum sustainable metabolic rate due to swimming) (see Goolish, 1991). Fry (1971) developed a classification of the effects of elements in the environment on the physiology and behavior of fishes, grouped into categories of controlling, limiting, masking (regulatory), and directive factors. This approach is known as the “Fry Paradigm” (Fry, 1971; Webb, 1978a; Brett, 1979; Kerr, 1990). Because environmental factors that affect scope also affect maximum sustained and prolonged speed, the Fry Paradigm provides a framework in which to consider environmental effects on these indices of performance.

Controlling factors, temperature and pressure, determine rates of enzyme-catalyzed reactions by their effects on the kinetic energy of the substrates (Fry, 1971). Metabolic scope increases with temperature, and with it, maximum swimming speed (Beamish, 1978; Brett, 1979).

The maximal limits to achievable performance with any given quantities of controlling factors are only possible in the absence of limiting factors. These typically reduce active metabolism by interfering with the delivery of oxygen and substrates, and the removal of wastes. There are often thresholds below which limiting factors reduce swimming performance, reflecting species variation in ability to regulate metabolite delivery and removal.

Many factors reduce water oxygen concentration, including altitude, dissolved substances (e.g., salt in seawater), suspended material (e.g., silt, fiber), and temperature. Active metabolic rate and swimming performance typically peak at a temperature below the upper limit of the normal tolerance range for a species. This has been attributed to a limiting interaction of temperature reducing water oxygen concentration (Brett, 1964). The same peaked relationship between active metabolic rate and temperature also can be found for terrestrial ectotherms (see Gordon, 1977), so it is difficult to accept this explanation.

Any environmental factor that perturbs a regulated system will require the expenditure of metabolic energy for its regulation. Fry (1971) called these masking factors, but regulatory factors is more descriptive (Webb, 1978a). Regulatory factors include pH, salinity, toxicants, and, for endothermic fishes, temperature. These factors increase standard metabolic rates and lower metabolic scope. This suggests that swimming performance, which is easily measured, would be an effective and economical method for quantifying environmental stress. However, when there are competing oxygen demands, the less immediate demand may be deferred. The stimulus to swim in experiments is compelling, and fish may defer regulatory costs to maximize locomotor performance. As a result, u_{crit} may not be reduced (Skadsen *et al.*, 1980). Fry (1971) called this phenomenon a modulated response.

Directive factors, among other effects, alter physiological profiles in time, for example, in seasonal acclimation, ontogenetic changes with niche shifts (e.g., alternation of freshwater and marine stages in salmonids and eels), and sexual maturation. Metabolic scope may be changed by these events as a result of changes at the enzyme–substrate level (e.g., Johnston and Harrison, 1985) and metabolite delivery systems (Cech *et al.*, 1976).

Routine swimming performance is presumed to be affected by energetic factors. One measure of optimal performance is the optimum cruising speed, defined as the speed at which the energy expenditure to move a unit mass through a unit distance is minimized (Weihs, 1973c). Similarly, foraging is often analyzed using contingency models, in which speed determines the rate of encounter with food particles (Weihs, 1975; Ware, 1975, 1978; Werner, 1986). Because energy costs of feeding increase faster than food consumption, there is an optimal foraging speed at which surplus energy is maximized per unit search cost. The optimal cruising speed is close to the speed at which activity metabolism is twice the standard metabolism, while optimal foraging speeds are slightly higher (Weihs and Webb, 1983). It follows that these speeds increase with fish size and with environmental temperature. Similarly, stressors that increase standard metabolic rates will increase optimal speeds for foraging.

There is a minimum speed during routine swimming that enables negatively buoyant fish to use hydrodynamic lift to support the body weight in water (Magnuson, 1978). This speed varies among species and depends on body density, morphology of the pectoral fins, and body and caudal peduncle keel, all of which generate lift (Magnuson, 1978). Minimum speeds for thunnids range from 0.3 to 2.2 $L\ s^{-1}$ (Magnuson, 1978).

5. *Sprints*

Sprints are short-duration bouts of high-level activity supported by anaerobic metabolism, with zero linear and centripetal acceleration. In practice, there are relatively few reliable measurements of maximum sprint speed, u_{max}, largely because a quantifiable sustained stimulus is lacking that ensures maximum performance (Wood, 1991). Although experimental systems try to place fish in situations where they are likely to perform maximally, reported u_{max} are mainly voluntary speeds. Training fish to sprint in response to a suitable stimulus appears to have the greatest promise for measuring u_{max} (Wardle and

Kanwisher, 1974; Videler and Wardle, 1978; Videler, 1981; Videler and Hess, 1984; Wardle and He, 1988). Observations in predator–prey situations might be expected to force maximum performance from the participants, but in practice u_{max} is rarely attained (Webb, 1981; Webb and Carolla, 1981). Predator speeds are typically submaximal, probably because prey escape response thresholds increase rapidly with predator attack speed (Dill, 1974; Webb, 1984b). Prey are not often chased far (Major, 1978), as a result of which prey responses are usually sufficient only to escape the predator's attack trajectory, and the duration of the response is too short to reach u_{max} (Webb and Carolla, 1981). Other methods startle fish in various ways but, because of methodological differences, it is difficult to compare results with confidence.

Sprint speeds vary over 1.5 orders of magnitude, from about 1.3 to 22 $L\ s^{-1}$ (Beamish 1978; Videler and Wardle, 1991). The Q_{10} value for maximum tail-beat frequency, and presumably u_{max}, is about 2 (Videler and Wardle, 1991). Morphological factors, such as propulsor shape, muscle mass, and muscle trajectories, undoubtedly affect sprint performance (Lighthill, 1975; McLaughlin and Kramer, 1991; Alexander, 1969). On average, maximum sprint speed, u_{max} (cm s^{-1}), increases with size, L (cm), according to (Videler and Wardle, 1991)

$$u_{max} = 40 + (7.4\ L). \tag{14}$$

Because of the technical difficulty of directly measuring u_{max}, indirect methods are popular. These usually compare muscle performance limits with drag, solving Eq. (1) for u (e.g., Bainbridge, 1961; Somero and Childress, 1980; Johnsrude and Webb, 1985; Goolish, 1991). Wardle (1975) proposed a different approach, arguing that maximum speed is determined by the minimum twitch time of myotomal muscle and fish stride length. The original supporting evidence of tail-beats recorded on video was not strong because imaging rates were slow compared to tail-beat periods. A 35-cm mackerel (*Scomber scombrus*) has been shown to sprint with a tail-beat frequency of 18 Hz, close to the expected value of 19 Hz for the species (Wardle and He, 1988). Wardle's method is attractively simple, but more comparative data on speed maxima and muscle characteristics are essential to fully evaluate it.

Sprints traditionally have been defined as speeds maintained for up to 15 s, but are probably much shorter in nature. Maximum sprints are probably rare events that occur only in fitness-critical situations, such as prey capture, predator evasion, reproduction, and reproductive migrations (Brett, 1983). Prey fish are typically within a few meters of cover provided by habitat structure (Hobson, 1974), other fish in schools (Major, 1978), or even in the background light space, particularly when in turbid water (Lythgoe, 1979). Cover can usually be reached in 3 or 4 s (Webb, 1986a). Lunges by predators at fish prey are even shorter in duration, typically lasting less than a second (Webb, 1986b; Puckett and Dill, 1984, 1985). Salmonids traverse high flows, including waterfalls, in seconds (Gray, 1968). Bed digging by female salmonids may take longer, but the metabolic activity required is unknown (Beacham, 1985). In general, it appears that sprints and other bursts should be measured for activities lasting less than 5 s.

B. Acceleration

1. Acceleration in Routine Swimming

Like swimming speed, linear and centripetal acceleration rates are small during routine swimming, but they rarely have been measured (Nursall, 1958; Webb, 1991). Median and paired fins are used by fish to rotate without translation ($u = 0$; $du/dt = 0$) (Blake, 1978). Maneuvers with translation are powered by BCF or MPF propulsors. The momentum in a glide may be used to maneuver, using fins to act as asymmetrical brakes or rudders (Aleyev, 1977).

Maneuvers in routine swimming appear to increase mechanical energy requirements by a factor of about 3, compared to that expected for net resistance from vector addition of the resistance components [Eq. (5)] (Weatherley *et al.*, 1982; Webb, 1991). This may be due to low efficiency, increased drag due to nonpropulsive use of fins, or unknown interference effects (Webb, 1991). Therefore, although acceleration rates are small, the costs of maneuvering are important contributors to the routine locomotor energy budget. As a result, optimal speeds estimated from rates of oxygen consumption may overestimate performance, whereas estimates of routine metabolic energy expenditure based on mean speeds in routine swimming may be underestimates.

2. Intermittent Swimming

Weihs (1973a, 1974) showed that intermittent swimming could provide substantial energetic advantages when fish had to swim continuously. The few empirical observations suggest that intermittent swimming is used when fatigue may occur, either at very low speeds where efficiency would be low (Rome *et al.*, 1990) or at high speeds where power requirements are high (Videler and Weihs, 1982). The onset of intermittent swimming and the proportion of time for which it is used vary among species (Fig. 7). Strong fusiform swimmers begin using burst-and-coast swimming at speeds of about half the 2 min u_{crit} (Fig.

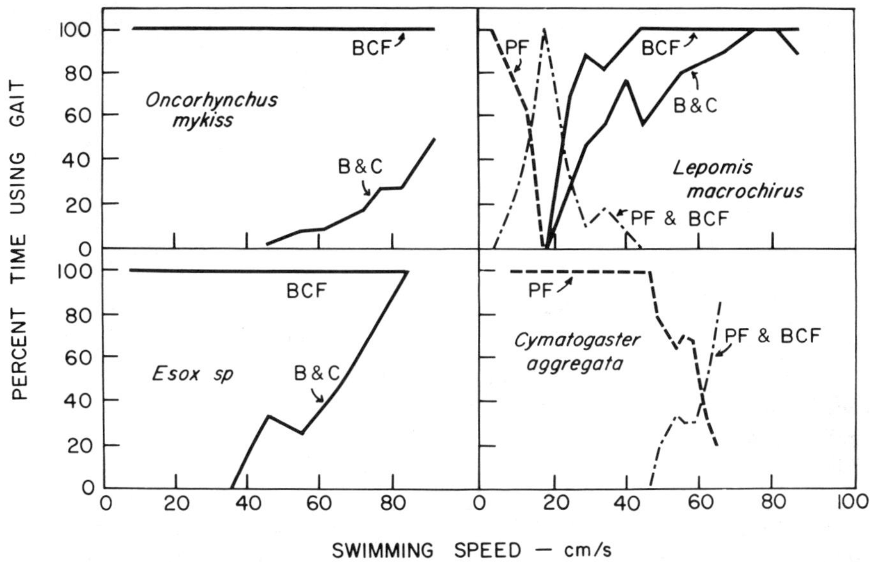

FIG. 7. The use of burst-and-coast gaits at sustained and prolonged speeds by four species of fish. Trout (*Oncorhynchus mykiss*) is a strong cruiser, whereas tiger musky (*Esox* sp.) is a sprinter. The shiner surfperch (*Cymatogaster aggregata*) is a strong labriform swimmer. Bluegill (*Lepomis macrochirus*) is a typical acanthopterygian found in structurally complex habitats. Data from P. W. Webb, in Alexander (1989) and Webb (1973). BCF, body/caudal fin swimming; B & C, burst and coast swimming; PF, pectoral fin swimming.

7), and continue to use this behavior at low sprint speeds on the order of 20–50% of the maximum sprint speed (Videler and Weihs, 1982).

Intermittent swimming may not be used continuously. When first used at low swimming speeds, burst-and-coast bouts are rare, but the period of time spent using this behavior increases with mean speed (Fig. 7). Intermittent swimming is probably used continuously in the low sprint speed range.

3. *Fast-Starts*

Fast-starts are maximum accelerations from rest and are used by prey escaping predator attacks and by some predators making those attacks (Eaton *et al.*, 1977; Webb and Skadsen, 1980; Harper and Blake, 1991). By convention, the fast-starts portion of a swimming burst is restricted to the first two tail sweeps, lasting a total of 50–200 ms. Subsequently, behavior becomes variable as fish continue moving and use a wide range of behaviors, from sprinting to braking.

Maximum acceleration rate has proved difficult to measure for several reasons, and results should be viewed with caution. First, fast-start performance has usually been reported for linear acceleration along the path of the center of mass. The effects of turning on linear acceleration performance have been neglected [Eq. (5)].

Second, early studies estimated acceleration rate from double differentiation of distance–time data (Weihs, 1973b; Webb, 1975b). It is well known that this approach oversmooths (Eaton *et al.*, 1977), an effect that Harper and Blake (1989) suggest may result in underestimates of maximum accelerations by as much as 30%. The problems of oversmoothing have led Harper and Blake (1989) to advocate the use of implanted accelerometers to measure maximum acceleration rates, a method pioneered by DuBois *et al.* (1976). This method is restricted in use to large fish, and also involves a third problem: unidirectional accelerometer measurements require corrections for turning. The correction is calculated by differentiation of distance–time data from high-speed film synchronized with the accelerometer record. Although the smoothing error is not as severe as for double differentiation, the correction will still be an underestimate (Harper and Blake, 1989). Therefore, resultant values for linear acceleration will be overestimates, as the correction is subtracted from the accelerometer reading. In addition, accelerations are relevant to the center of mass that, in Harper and Blake's studies, was not measured but assumed from the literature.

Although maximum acceleration performance is not known with certainty, mean performance can be accurately measured as the distance traveled with time and turning angles (e.g., Eaton and Emberley, 1991). Neurophysiologists interested in control problems adopted this approach early because it is especially useful for differentiating among fast-start patterns (e.g., Eaton and Emberley, 1991). Indirect measures are similarly useful for mechanical studies (Webb, 1978b; Webb *et al.*, 1991, 1992).

Results from various studies of fast-starts have been summarized by Harper and Blake (1990) and Domenici and Blake (1991). Fish have impressive abilities for linear acceleration, with maximum rates on the order of 2–16*g*, and average rates of 0.8–5.5*g* (Domenici and Blake, 1991). Early in ontogeny, acceleration performance increases with fish size (Webb and Carolla, 1981; Fuiman, 1989). Performance is little affected by size for intermediate sizes (Webb, 1976), and probably decreases for large fishes (Webb and de Buffrénil, 1990). Temperature has little effect on performance over much of the temperature tolerance range (Webb, 1978c).

Species effects on acceleration performance are unclear. Some studies have shown little or no difference in mean performance among species (Webb, 1978b), whereas others have shown the opposite (Webb, 1975b, 1986b; Harper and Blake, 1990). Large species variation in peak acceleration rates has been observed with accelerometry (Harper and Blake, 1990, 1991). In contrast, mean performance, such as the distance traveled in various times, appears less variable (Domenici and Blake, 1991). This may be because fish with large peak accelerations often also have large decelerations.

There is clearly some debate as to how acceleration performance should be measured and how results using different approaches relate to effects of functional morphology on performance. Solutions to this controversy probably relate to the difficult problem of ultimate causation. For example, if a sudden movement is important in deterring predators, perhaps by showing that potential prey are alert, maximizing peak acceleration may be the most important measure of fast-start performance. Alternatively, if displacement from predator strike trajectories is important, then average performance, such as distance traveled in a given time, might be a more appropriate measure of performance. The latter hypothesis is attractive because the similarity in average performance among species suggests that performance may be conserved in spite of morphological variation. Convergence in acceleration performance might be expected, as virtually all fish are prey at some stage during ontogeny. Measurements of acceleration performance, using various methods for members of clades with variable dependence on fast-starts, appear to be essential for resolving this issue (Brooks and McLennan, 1991; Feder *et al.*, 1987).

Studies of fast-starts by researchers interested in different questions and using different methods have spawned a confusing terminology, much of which overlaps (Table IV). Much of the terminology appears to be derived from details of responses that probably have common underlying physical causes. Therefore, to better understand the limits to fast-start performance, it is desirable to integrate the various data within a common theoretical framework. Foreman and Eaton (1993) have reached the same conclusion based on the study of neural control and fast-start behavior. I suggest that a unifying theme is control of turning and recoil of the anterior of the body, the latter being due to unbalanced side forces primarily from the caudal region of the body. Thus, fast-start performance must be considered in terms of simultaneous changes in both linear and angular momenta, measured as linear and centripetal accelerations (Daniel and Meyhofer, 1989; Domenici and Blake, 1991).

To illustrate the importance of control movements in fast-starts, consider as a reference a C-start with continued minimum radius turn-

TABLE IV

RECENT NOMENCLATURE USED TO DESCRIBE FAST-STARTS OF FISHES[a]

Description	Nomenclature	Source
Kinematics		
Stage 1		
Short latency, usually unilateral contraction of the muscle of the trunk and tail causing the body and tail to bend from a stretched-straight posture into a C-, J-, or S-shape; this movement may be omitted if the fish body is already curved into C-, J-, or S-shapes	Preparatory stage Fast-body-bend Stage 1	Weihs (1973b) Eaton *et al.* (1977) Webb (1975b)
Stage 2		
Tail moves almost perpendicular to the heading of the body as the body bends formed in Stage 1 travel caudally; the body ends in a C-, J-, or S-shape posture oriented in the opposite direction to that of Stage 1	Propulsive stage Return-flip Main propulsive stage Stage 2	Weihs (1973b) Eaton *et al.* (1977) Webb (1975b) Webb (1975b)
Stage 3		
Fish behavior varies from braking to a stop, through gliding, to establishing a propagating wave in continuing powered swimming	Final stage Stage 3 Variable stage	Weihs (1973b) Webb (1975b) Harper and Blake (1990)
Fast-start patterns		
Body bends into a C-shape during Stage 1	C-start	Eaton *et al.* (1977)
Body bends into a J-shape during Stage 1	J-start	Weihs (1973b)
Body bends into an S-shape during Stage 1	S-start Pattern A	Webb (1976) Webb and Skadsen (1980)
Before the fast-start begins, the body is bent into a shape approaching that seen at the end of a normal Stage 1	Pattern B	Webb and Skadsen (1980)
The head begins to lead a turn during Stage 1 that is in the opposite direction to that originally initiated during Stage 1	J-start Type III fast-start	Weihs (1973b) Harper and Blake (1990, 1991)

(continued)

TABLE IV (*Continued*)

Description	Nomenclature	Source
The head begins to lead a turn during Stage 2 that is in the opposite direction to the turn initiated as part of Stage 1	Counterturn Type II fast-start	Webb (1976) Harper and Blake (1990, 1991)
Performance patterns		
High linear acceleration and high unidirectional centripetal acceleration rates, resulting in single peak in acceleration time trace	Fast forward C-start Type I fast-start	Eaton *et al.* (1977) Harper and Blake (1990, 1991)
Two acceleration peaks during Stages 1 and 2	Type II fast-start	Harper and Blake (1990, 1991)
Three acceleration peaks during Stages 1 and 2	Type III fast-start	Harper and Blake (1990, 1991)
Four acceleration peaks during Stages 1 and 2	Type IV fast-start	Harper and Blake (1990, 1991)

[a] Headings are terms that appear to be in most common usage. In addition to first sources for various terms, descriptions draw on Eaton *et al.* (1977), Eaton and Emberley (1991), Webb (1976), Harper and Blake (1990), Vinyard (1982), and Domenici and Blake (1991).

ing in a constant direction (Eaton and Emberley, 1991). The C-start is named after the body shape seen during the first tail sweep (Stage 1). During Stage 1, propulsive elements move laterally at large angles to their axes of motion, and hence at a large angle to a line joining segments and the center of mass (Weihs, 1972, 1973b). As a result, normal forces generate large turning moments about the center of mass, with relatively little force accelerating the center of mass forward. Hence, Stage 1 of a C-start is an acceleration turn, with the net propulsive force increasing linear and angular momenta. The angle subtended by a propulsive element, relative to its axis of motion, decreases with time and becomes small during the second tail sweep (Stage 2). As a result, the component of the normal force acting through the center of mass increases, while the turning moment decreases (Weihs, 1973b).

The temporal pattern of the thrust force component through the center of mass is the basis for Weihs' (1973b) considering Stage 1 to be "preparatory," setting up a body posture to maximize the linear acceleration component during a "main propulsive" Stage 2. Eaton's term of "fast-body-bend" recognizes the importance of setting up a posture quickly in order to subsequently maximize the displacement of the center of mass during the "return flip" (Table IV).

Forces vary with time during Stages 1 and 2, just as they do during periodic tail-beats in steady swimming (DuBois *et al.*, 1976; Freadman, 1981; Videler, 1981; Harper and Blake, 1990). Thrust and acceleration increase at the start of each tail sweep, reach a maximum, and then decrease. At the end of a tail sweep, thrust is momentarily smaller than resistance, so fish decelerate. Linear acceleration patterns measured by accelerometry during such a C-start show a small peak in Stage 1, followed by a second, larger peak in Stage 2 (Fig. 8A). Harper and Blake (1990) called this acceleration pattern a Type I fast-start, because there is one major acceleration peak (in Stage 2). However, their Type I fast-start is seen to be a typical example of a C-start, the earlier and more descriptive term (Fig. 8A).

Consider departures from the reference C-start. One deviation is a decrease in the duration of the turn (lower turning angle) associated with an earlier increase in turning radius. Because resistance to centripetal acceleration is proportional to $1/r$, a larger turning radius would be expected to reduce angular acceleration resistance. Therefore, less energy should be needed for a fast-start with a smaller turning angle. For a given amount of energy, the reduction should lead to greater linear acceleration [Eq. (5)]. In practice, no such advantage is seen. Instead, performance is little affected by the nature of a startle response maneuver (Webb, 1976; Harper and Blake, 1990). One possible explanation for this is that fewer motor units are used for larger-radius turns (Eaton *et al.*, 1988; Foreman and Eaton, 1990). In this situation less total work would be done, but because less energy is dissipated in centripetal acceleration, linear acceleration performance would not be enhanced. The recruitment of a variable number of motor units could also explain variation in the magnitude of initial linear acceleration peaks that was shown by Harper and Blake (1990, 1991).

Varying the number of motor units can affect the initial turning angle, but cannot reverse the direction of a turn. Because some recoil turning must always occur, escape trajectories with angles greater than about 40° to the original body axis must result from generation of forces opposite to those produced by the initial muscle contraction. This force could result from advancing the time of onset of contraction of contralateral muscles, banking, use of paired fins, etc., to counteract or reverse the initial turn. Energy expended to reduce or reverse the growth of angular momentum would not be available to improve linear

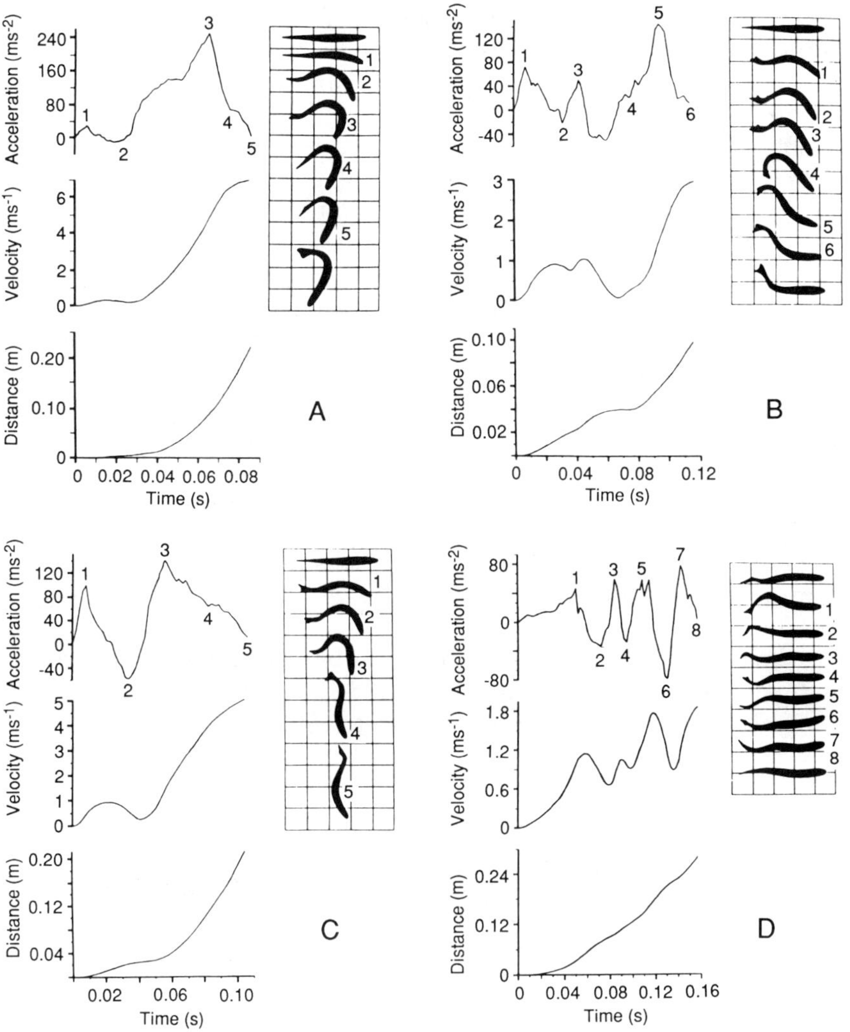

FIG. 8. Fast-starts and linear acceleration records obtained by unidirectional accelerometers for northern pike (*Esox lucius*) during forced fast-starts (A–C) and a strike at tethered prey (D). The numerals on the graphs match the silhouette numbers in the boxed insets showing kinematics. Figures A–C are from Harper and Blake (1990) and D is from Harper and Blake (1991).

acceleration, so again, apparent benefits for linear acceleration from reducing net angular momentum would not be realized.

The effect of working to reduce the turning radius of a C-start is to initiate a counterturn (Webb, 1976). Because decisions on the number of motor units to use are made before acceleration begins (Eaton and Emberley, 1991), and Stage 2 EMG can begin as early as Stage 1 movements (Eaton *et al.*, 1988; Eaton and Emberley, 1991), counterturns may be initiated early in a fast-start.

Harper and Blake (1990) describe as Type III those fast-starts with three major linear acceleration peaks. These appear to be counterturns beginning during Stage 1, which, not surprisingly, are associated with breaks in the acceleration (Fig. 8B). A Type II fast-start, defined by Harper and Blake as having two major acceleration peaks (Fig. 8C), appears to be a turn initiated during Stage 2. Type IV fast-starts (Fig. 8D), with four acceleration peaks, approach S-starts, with the body bent into S-shaped postures during Stage 1. Fast-starts with many more peaks appear to be feasible.

The simplest and most descriptive terms for various fast-start patterns are based on body postures in Stage 1 and are likened to C-, J-, and S-shapes. C-starts are seen to be sustained unidirectional turns. The J-shape (Weihs, 1973b) appears to be associated with direction changes initiated during Stage 1. In an S-start, the net turning (recoil) moment is reduced by a double bend in the body, but is not eliminated. As a result, the center of mass turns during Stage 1, but the turn is reversed in Stage 2 to achieve an overall rectilinear advance. Thus, S-starts result from turning control movements initiated before the fast-start begins, a final step in the timing of the onset of counterturn control. However, S-starts appear to require additional control counterturns in Stage 2 to prevent a recoil turn later in the fast-start.

The effect of control movements on linear acceleration performance is further illustrated by observations of pike during strikes at prey. Harper and Blake (1991) found that the number of acceleration peaks increased with the distance from a pike to its prey, even though the body shapes during strikes showed relatively little variation, compared to that shown in Fig. 8. Thus, directional control movements increased in frequency as strike distance increased, which is only to be expected. However, the control movements required to ensure striking the prey resulted in reduced mean and maximum linear acceleration rates.

Published data are inadequate to further test the ideas expressed here, but the pattern is consistent with existing data. The approach explains similarities in mean translation of the center of mass for maneuvers of different turning radius. It places observed variation in kinematics from C-starts through J-starts to S-starts on a continuum, in terms of common principles. These ideas are consistent with patterns of linear acceleration peaks (Harper and Blake, 1990, 1991; Domenici and Blake, 1991) and with the performance of pike in prey strikes (Harper and Blake, 1991).

In general, therefore, the observed variation in acceleration rates and kinematic patterns appears to be explainable by the intermittent nature of the thrust force during tail sweeps (DuBois *et al.*, 1976; Freadman, 1981; Foreman and Eaton, 1993) and dissipation of energy

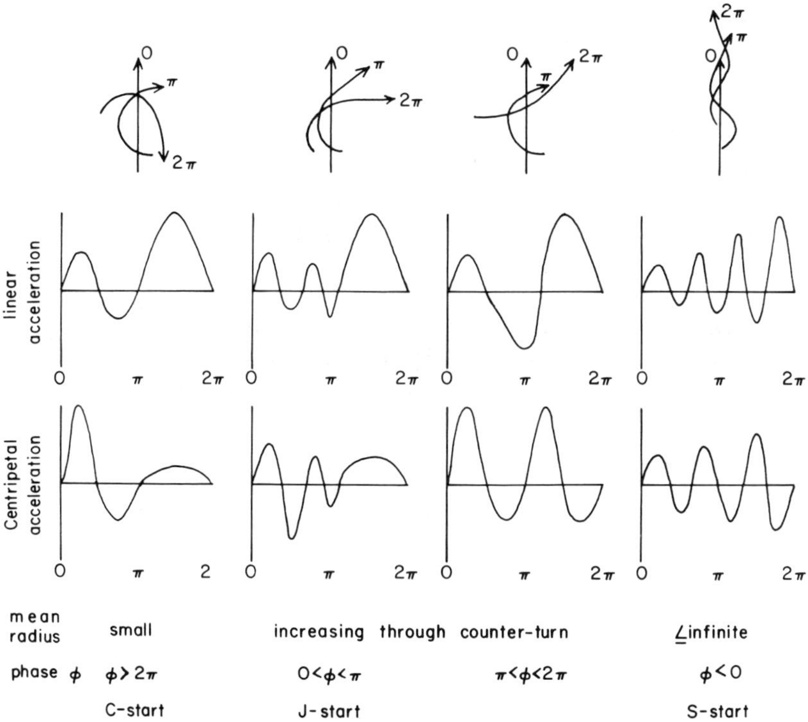

FIG. 9. Schematic representation of fast-start patterns described for fish. These illustrate common features and suggest a general descriptive approach to performance that recognizes the importance of both linear and turning accelerations, and both the timing and magnitude of control movements. Generic linear and centripetal acceleration patterns are suggested based on observations in DuBois *et al.* (1976), Freadman (1981), Harper and Blake (1990, 1991), Domenici and Blake (1991), and P. W. Webb (unpublished biaxial accelerometer observations). The reference C-start is based on Eaton and Emberley (1991). Other tracings are based on Weihs (1972, 1973b), Webb (1975b, 1976, 1978b), Harper and Blake (1990, 1991), and Domenici and Blake (1991).

in turning control movements. However, rather than being separate "types" of acceleration, there is variation in timing of a common feature: the onset of directional control motions. Thus, virtually any pattern and number of acceleration peaks is possible. Therefore, I suggest performance in fast-starts could be described by three variables: linear acceleration, angular acceleration, and the phase relationship for the onset of changes in direction (Fig. 9). A similar conclusion has been reached by Foreman (1991) from observations of EMG and kinematics. This suggestion has parallels with muscle work loops, for which numerous contraction patterns arise from the interplay of a few variables (Josephson, 1985). Muscle physiologists have not burdened themselves with names for different loop shapes, and by analogy, perhaps the burgeoning terminology for fast-starts could be reduced or eliminated.

4. Turning

Turning circles of fish can have very small radii, typically less than 20% of body length (Webb, 1983; Domenici and Blake, 1991). Howland (1974) suggested that radius should be independent of speed but proportional to the projected area of the body generating turning moments. These predictions have been supported for fusiform fish (Webb, 1983). However, the minimum turning radius of angelfish (*Pterophyllum*) is smaller than predicted by Howland's model (Domenici and Blake, 1991). These fish are compressed and probably have greater flexibility and, hence, ability to generate larger turning moments (Domenici and Blake, 1991). Howland's model requires modification to incorporate these additional effects, while observations on turning performance are required for a large range of species differing in body and fin form.

VI. Conclusions

Motor performance of fish should be measured in terms of speed, u, linear acceleration rate, du/dt, and turning radius, r, in as many planes as are relevant to a given problem. Published performance limits implicitly assume that one or more of these co-varying performance parameters is negligible, so reported performance maxima should be viewed with caution. Motor performance measures other than speed are rarely measured during submaximal activity, but may substantially add to rates of energy dissipation. Experimental studies with a more pluralistic and mechanically rigorous approach to the interacting factors affecting motor and physiological performance are required.

Traditional analyses of fish locomotion have focused on variation in propulsor use among phylogenetic groups and recruitment patterns of myotomal muscle with increasing speed. Intraspecific locomotor patterns have been neglected. Present knowledge of propulsors and the need to efficiently meet energy requirements over a very large performance range can be integrated following Alexander's (1989) application of the concept of gaits during swimming. This concept shows that fish use a multigear system involving combinations of discrete muscles, discrete propulsors, and various behaviors in a series of gaits that finely divide the range of swimming performance (Webb, 1994).

Performance is usually classified on the basis of endurance–speed relationships of strong BCF swimmers. The form of the relationship over the whole performance range for fish is unknown, and different performance levels are based on predetermined endurance times. An alternative view is suggested in which gait changes play a large role in determining the relationship. This approach requires greater attention to changes in the use of various muscle, propulsor, and behavior combinations with increasing performance. It also provides a framework within which correlations among morphologies, gaits, and performance capabilities may lead to greater understanding of evolutionary and ecological patterns among fishes. Data on performance limits in various gaits is fragmentary, except for BCF gaits, in large part because the idea of using gait concepts for locomotion in fluids is very new (Alexander, 1989).

Although BCF prolonged and sprint speeds, and fast-start accelerations, have been studied for some time, their performance limits are not clear. Methods, values, and significance remain topics of debate. In particular, studies of fast-starts require analyses using more rigorous biomechanical criteria to seek common principles.

Finally, very little is known about performance limits and how they are determined for most MPF gaits. This is an oversight, because these propulsor systems are probably the dominant mechanisms for routine locomotion. Hence, knowledge of MPF swimming is likely to provide essential insights for understanding habitat choices and habits of many fish species.

Acknowledgments

This analysis draws on research supported by the National Science Foundation and was completed with support from NSF Grant DCB9017817.

References

Alexander, R. McN. (1967). "Functional Design of Fishes." Hutchinson, London.

Alexander, R. McN. (1969). Orientation of muscle fibres in the myomeres of fishes. *J. Mar. Biol. Assoc. U.K.* **49,** 263–290.

Alexander, R. McN. (1989). Optimization and gaits in the locomotion of vertebrates. *Physiol. Rev.* **69,** 1199–1227.

Alexander, R. McN. (1990). Size, speed and buoyancy adaptations in aquatic animals. *Am. Zool.* **30,** 189–196.

Alexander, R. McN. (1991). Energy-saving mechanisms in walking and running. *J. Exp. Biol.* **160,** 55–69.

Aleyev, Y. G. (1977). "Nekton." Junk Publ., The Hague.

Archer, S. D., and Johnston, I. A. (1991). Kinematics of labriform and subcarangiform swimming in the Antarctic fish, *Notothenia neglecta*. *J. Exp. Biol.* **143,** 195–210.

Arnold, G. P., and Weihs, D. (1978). The hydrodynamics of rheotaxis in the plaice (*Pleuronectes platessa*). *J. Exp. Biol.* **75,** 147–169.

Arnold, G. P., Webb, P. W., and Holford, B. H. (1991). The role of the pectoral fins in station-holding of Atlantic salmon (*Salmo salar* L). *J. Exp. Biol.* **156,** 625–629.

Bainbridge, R. (1961). Problems of fish locomotion. *Symp. Zool. Soc. London* **40,** 23–56.

Bainbridge, R. (1962). Training, speed and stamina in trout. *J. Exp. Biol.* **39,** 537–555.

Beacham, T. D. (1985). Meristic and morphometric variation in pink salmon (*Oncorhynchus gorbuscha*) in southern British Columbia and Puget Sound. *Can. J. Zool.* **63,** 366–372.

Beamish, F. W. H. (1978). Swimming capacity. *In* "Fish Physiology" (W. S. Hoar and D. J. Randall, eds.), Vol. 7, pp. 101–187. Academic Press, Orlando, FL.

Bennett, A. F. (1980). The metabolic foundations of vertebrate behavior. *BioScience* **30,** 452–456.

Bennett, A. F. (1991). The evolution of activity capacity. *J. Exp. Biol.* **160,** 1–23.

Blake, R. W. (1977). On ostraciiform locomotion. *J. Mar. Biol. Assoc. U.K.* **57,** 1047–1055.

Blake, R. W. (1978). On balistiform locomotion. *J. Mar. Biol. Assoc. U.K.* **58,** 73–80.

Blake, R. W. (1979). The energetics of hovering in the mandarin fish (*Synchropus picturatus*). *J. Exp. Biol.* **82,** 25–33.

Blake, R. W. (1980). Undulatory median fin propulsion of two teleosts with different modes of life. *Can. J. Zool.* **58,** 2116–2119.

Blake, R. W. (1981). Mechanics of ostraciiform propulsion. *Can. J. Zool.* **59,** 1067–1071.

Blake, R. W. (1983a). "Fish Locomotion." Cambridge Univ. Press, Cambridge, UK.

Blake, R. W. (1983b). Swimming in the electric eels and knifefishes. *Can. J. Zool.* **61,** 1432–1441.

Bone, Q. (1966). On the function of the two types of muscle fibre in elasmobranch fish. *J. Mar. Biol. Assoc. U.K.* **46,** 321–349.

Bone, Q. (1977). Muscular and energetic aspects of fish swimming. *In* "Swimming and Flying in Nature" (Y. T. Wu, C. J. Brokaw, and C. Brennen, eds.), pp. 493–528. Plenum, New York.

Braun, J., and Reif, W. (1985). A survey of aquatic locomotion in fishes and tetrapods. *Neues. Jahrb. Geol. Palaeontol., Abh.* **169,** 307–332.

Breder, C. M. (1926). The locomotion of fishes. *Zoologica (N.Y.)* **4,** 159–256.

Brett, J. R. (1964). The respiratory metabolism and swimming performance of young sockeye salmon. *J. Fish. Res. Board Can.* **21,** 1183–1226.

Brett, J. R. (1965). The relation of size to the rate of oxygen consumption and sustained swimming speeds of sockeye salmon (*Oncorhynchus nerka*). *J. Fish. Res. Board Can.* **22,** 1491–1501.

Brett, J. R. (1967). Swimming performance of sockeye salmon in relation to fatigue time and temperature. *J. Fish. Res. Board Can.* **24,** 1731–1741.

Brett, J. R. (1979). Environmental factors and growth. *In* "Fish Physiology" (W. S. Hoar and D. J. Randall, eds.), Vol. 8, pp. 599–675. Academic Press, Orlando, FL.

Brett, J. R. (1983). Life energetics of sockeye salmon, *Oncorhynchus nerka. In* "Behavioral Energetics" (W. P. Aspey and S. I. Lustick, eds.), pp. 29–63. Ohio State Univ. Press, Columbus.

Brett, J. R., and Sutherland, D. B. (1965). Respiratory metabolism of pumpkinseed, *Lepomis gibbosus*, in relation to swimming speed. *J. Fish. Res. Board Can.* **22,** 405–409.

Brooks, D. R., and McLennan, D. A. (1991). "Phylogeny, Ecology, and Behavior." Univ. of Chicago Press, Chicago.

Cech, J. J., Bridges, D. M., Rowell, D. M., and Balzer, P. J. (1976). Cardiovascular responses of winter flounder *Pseudopleuronectes americanus* (Walbaum) to acute temperature increase. *Can. J. Zool.* **54,** 1383–1388.

Daniel, T. L. (1984). Unsteady aspects of aquatic locomotion. *Am. Zool.* **24,** 121–134.

Daniel, T. L. (1988). Forward flapping flight from flexible fins. *Can. J. Zool.* **66,** 630–638.

Daniel, T. L., and Meyhofer, E. (1989). Size limits in escape locomotion of Carridean shrimp. *J. Exp. Biol.* **143,** 245–265.

Daniel, T. L., and Webb, P. W. (1987). Physics, design and locomotor performance. *In* "Comparative Physiology: Life in Water and on Land" (P. Dejours, L. Bolis, C. R. Taylor, and E. R. Weibel, eds.), pp. 343–369. Springer-Verlag, New York.

Davison, W., and MacDonald, J. A. (1985). A histochemical study of the swimming musculature of antarctic fish. *N. Z. J. Zool.* **12,** 473–483.

Dill, L. M. (1974). The escape response of the zebra danio (*Brachydanio rerio*). I. The stimulus for escape. *Anim. Behav.* **22,** 711–722.

Domenici, P., and Blake, R. W. (1991). The kinematics and performance of the escape in the angelfish (*Pterophyllum eimekei*). *J. Exp. Biol.* **156,** 187–205.

Dorn, P., Johnson, L., and Darby, C. (1979). The swimming performance of nine species of common California inshore fishes. *Trans. Am. Fish. Soc.* **108,** 366–372.

DuBois, A. B., Cavagna, G. A., and Fox, R. S. (1976). Locomotion of bluefish. *J. Exp. Zool.* **195,** 223–236.

Eaton, R. C., and Emberley, D. S. (1991). How stimulus direction determines the trajectory of the Mauthner-initiated escape response in a teleost fish. *J. Exp. Biol.* **161,** 469–487.

Eaton, R. C., Bombardieri, R. A., and Meyer, D. L. (1977). The Mauthner-initiated startle response in teleost fish. *J. Exp. Biol.* **66,** 65–81.

Eaton, R. C., DiDomenico, R., and Nissanov, J. (1988). Flexible body dynamics of the goldfish C-start: Implications for reticulospinal command mechanisms. *J. Neurosci.* **8,** 2758–2768.

Eckert, R. (1988). "Animal Physiology." Freeman, New York.

Farlinger, S., and Beamish, F. W. H. (1977). Effects of time and velocity increments in the critical swimming speeds of largemouth bass (*Micropterus salmoides*). *Trans. Am. Fish. Soc.* **106,** 436–439.

Feder, M. E., Bennett, A. F., Burggren, W., and Huey, R. B. (1987). "New Directions in Ecological Physiology." Cambridge Univ. Press, Cambridge, UK.

Foreman, M. B. (1991). The kinematics and neurobiology of the Mauthner-initiated escape response. Ph.D. Dissertation, University of Colorado, Boulder.

Foreman, M. B., and Eaton, R. C. (1990). EMG and kinematic analysis of the stages of the Mauthner-initiated escape response. *Soc. Neurosci. Abstr.* **16,** 1328.

Foreman, M. A,. and Eaton, R. C .(1993). The directional change concept for reticulospinal control of goldfish escape. *J. Neurosci.* **13,** 4101–4113.

Freadman, M. A. (1981). Swimming energetics of striped bass (*Morone saxatilis*) and bluefish (*Pomatomus saltatrix*): Hydrodynamic correlates of locomotion and gill ventilation. *J. Exp. Biol.* **90,** 253–265.

Fry, F. E. J. (1971). The effect of environmental factors on the physiology of fish. *In* "Fish Physiology" (W. S. Hoar and D. J. Randall, eds.), Vol. 6, pp. 1–98. Academic Press, Orlando, FL.

Fuiman, L. A. (1989). Vulnerability of Atlantic herring larvae to predation by yearling herring. *Mar. Ecol.: Prog. Ser.* **51,** 291–299.

Fuiman, L. A., and Webb, P. W. (1988). Ontogeny of routine swimming activity and performance in zebra danios (Teleostei: Cyprinidae). *Anim. Behav.* **36,** 250–261.

Gee, J. H. (1983). Ecological implications of buoyancy control in fish. *In* "Fish Biomechanics" (P. W. Webb and D. Weihseds, eds.), pp. 140–176. Praeger, New York.

Goldspink, G. (1981). The use of muscles during flying, swimming, and running from the point of view of energy saving. *Symp. Zool. Soc. London* **48,** 219–238.

Goolish, E. M. (1991). Aerobic and anaerobic scaling in fish. *Biol. Rev. Cambridge Philos. Soc.* **66,** 33–56.

Gordon, M. S. (1977). "Animal Physiology." Macmillan, New York.

Gray, J. (1936). Studies in animal locomotion. VI. The propulsive powers of the dolphin. *J. Exp. Biol.* **13,** 192–199.

Gray, J. (1968). "Animal Locomotion." Weidenfeld & Nicolson, London, England.

Grobecker, D. B., and Pietsch, T. W. (1979). High-speed cinematographic evidence for ultrafast feeding in antennariid anglerfishes. *Science* **205,** 1161–1162.

Harper, D. G., and Blake, R. W. (1989). On the error involved in high-speed film when used to evaluate maximum acceleration of fish. *Can. J. Zool.* **67,** 1929–1936.

Harper, D. G., and Blake, R. W. (1990). Fast-start performance of rainbow trout *Salmo gairdneri* and northern pike *Esox lucius*. *J. Exp. Biol.* **150,** 321–342.

Harper, D. G., and Blake, R. W. (1991). Prey capture and the fast-start performance of northern pike *Esox lucius*. *J. Exp. Biol.* **155,** 175–192.

Harrison, P., Nicol, J. M., and Johnston, I. A (1987). Gross morphology, fibre composition and mechanical properties of pectoral fin muscles in the Antarctic Teleost, *Notothenia neglecta* Nybelin. *In* "Proceedings of the Fifth Congress of European Ichthyology" (K. O. Kulander and B. Fernholm, eds.), pp. 459–465. Stockholm, Sweden.

Hoar, W. S., and Randall, D. J., eds. (1978). "Fish Physiology" Vol. 7. Academic Press, Orlando, FL.

Hobson, E. S. (1974). Feeding relationships of teleostean fishes on coral reefs in Kona, Hawaii. *Fish. Bull.* **72,** 915–1031.

Hochachka, P. W., and Somero, G. N. (1984). "Biochemical Adaptation." Princeton Univ. Press, Princeton, NJ.

Hoerner, S. F. (1965). "Fluid-Dynamic Drag." Hoerner Fluid Dynamics, Brick Town, NJ.

Hoerner, S. F. (1975). "Fluid-Dynamic Lift." Hoerner Fluid Dynamics, Brick Town, NJ.

Howland, H. C. (1974). Optimal strategies for predator avoidance: The relative importance of speed and manoeuverability. *J. Theor. Biol.* **47,** 333–350.

Hynes, H. B. N. (1970). "The Ecology of Running Waters." Univ. of Toronto Press, Toronto.

Johnsrude, C. L., and Webb, P. W. (1985). Mechanical properties of the myotomal musculo-skeletal system of rainbow trout, *Salmo gairdneri*. *J. Exp. Biol.* **119,** 71–83.

Johnston, I. A. (1981). Structure and function of fish muscle. *Symp. Zool. Soc. London* **48,** 71–113.

Johnston, I. A. (1991). Muscle action during locomotion: A comparative perspective. *J. Exp. Biol.* **160,** 167–185.

Johnston, I. A., and Harrison, P. (1985). Contractile and metabolic characteristics of muscle fibers from antarctic fish. *J. Exp. Biol.* **116,** 223–236.

Johnston, I. A., Davison, W., and Goldspink, G. (1977). Energy metabolism of carp swimming muscle. *J. Comp. Physiol.* **114,** 203–216.

Jones, D. R. (1971). Theoretical analysis of factors which may limit the maximum oxygen uptake of fish: The oxygen cost of the cardiac and branchial pumps. *J. Theor. Biol.* **32,** 341–349.

Jones, D. R., and Randall, D. J. (1978). The respiratory and circulatory systems during exercise. *In* "Fish Physiology" (W. S. Hoar and D. J. Randall, eds.), Vol. 7, pp. 405–501. Academic Press, Orlando, FL.

Josephson, R. K. (1985). Mechanical power output from striated muscle during cyclic contractions. *J. Exp. Biol.* **114,** 493–512.

Kerr, S. R. (1990). The Fry Paradigm: Its significance for contemporary ecology. *Trans. Am. Fish. Soc.* **119,** 779–785.

Lagler, K. F., Bardach, J. E., Miller, R. R., and Pasino, D. R. M. (1977). "Ichthyology." Wiley, New York.

Lighthill, J. (1975). "Mathematical Biofluiddynamics." Soc. Ind. Appl. Math., Philadelphia.

Lighthill, J. (1990a). Biofluiddynamics of balistiform and gymnotiform locomotion. Part 2. The pressure distribution arising in two-dimensional irrotational flow from a general symmetrical motion of a flexible flat plate normal to itself. *J. Fluid Mech.* **213,** 1–10.

Lighthill, J. (1990b). Biofluiddynamics of balistiform and gymnotiform locomotion. Part 3. Momentum enhancement in the presence of a body of elliptical cross-section. *J. Fluid Mech.* **213,** 11–20.

Lighthill, J. (1990c). Biofluiddynamics of balistiform and gymnotiform locomotion. Part 4. Short-wavelength limitations on momentum enhancement. *J. Fluid Mech.* **213,** 21–32.

Lighthill, J., and Blake, R. W. (1989). Biofluiddynamics of balistiform and gymnotiform locomotion. Part 1. Biological background, and analysis by elongated-body theory. *J. Fluid Mech.* **212,** 183–207.

Lindsey, C. C. (1978). Form, function, and locomotory habits in fish. *In* "Fish Physiology" (W. S. Hoar and D. J. Randall, eds.), Vol. 7, pp. 1–100. Academic Press, Orlando FL.

Lythgoe, J. N. (1979). "The Ecology of Vision." Oxford Univ. Press (Clarendon), Oxford.

MacDonnell, A. J., and Blake, R. W. (1990). Rheotaxis in *Otocinclus* sp. (Teleostei: Loricariidae). *Can. J. Zool.* **68,** 599–601.

Magnuson, J. J. (1978) Locomotion by scombroid fishes: Hydromechanics, morphology, and behavior. *In* "Fish Physiology" (W. S. Hoar and D. J. Randall, eds.), Vol. 7, pp. 239–313. Academic Press, Orlando, FL.

Major, P. F. (1978). Predator–prey interactions in two schooling fishes, *Caranx ignobilis* and *Stolephorus purpureus*. *Anim. Behav.* **26,** 760–777.

Matthews, W. J. (1985). Critical current speeds and microhabitats of the benthic fishes *Percina roanoka* and *Etheostoma flabellare*. *Environ. Biol. Fishes* **12,** 303–308.

McCutcheon, C. W. (1971). Froude propulsive efficiency of a small fish, measured by wake visualization. *In* "Scale Effects in Animal Locomotion" (T. J. Pedley, ed.), pp. 339–363. Academic Press, Orlando, FL.

McLaughlin, R. L., and Kramer, D. L. (1991). The association between amount of red muscle and mobility in fishes: A statistical evaluation. *Environ. Biol. Fishes* **30,** 369–378.

Montgomery, A. R., and MacDonald, J. A. (1984). Performance of motor systems in antarctic fishes. *J. Comp. Physiol. A* **154A,** 241–248.

Moy-Thomas, J. A., and Miles, R. S. (1971). "Palaeozoic Fishes." Saunders, Philadelphia.

Nursall, J. R. (1958). A method of analysis of the swimming of fish. *Copeia*, pp. 136–141.

Nursall, J. R. (1973). Some behavioral interactions of spottail shiners (*Notropis hudsonius*), yellow perch (*Perca flavescens*), and northern pike (*Esox lucius*). *J. Fish. Res. Board Can.* **30,** 1161–1178.

O'Brien, W. J., Evans, B. I., and Howick, G. L. (1986). A new view of the predation cycle of a planktivorous fish, white crappie (*Pomoxis annularis*). *Can. J. Fish. Aquat. Sci.* **43,** 1894–1899.

Pavlov, D. S., and Tyurukov, S. N. (1988). The role of hydrodynamic stimuli in the behavior and orientation of fishes near obstacles. *Vopr. Ikhtiol.* **28,** 303–314.

Pavlov, D. S., Skorobagatov, M. A., and Shtaf, L. G. (1982). The critical current velocity of fish and the degree of flow turbulence. *Dokl. Akad. Nauk SSSR* **267,** 1019–1021.

Pavlov, D. S., Skorobagatov, M. A., and Shtaf, L. G. (1983). Threshold speeds for rheoreaction of roach in flows with different degrees of turbulence. *Dokl. Akad. Nauk SSSR* **268,** 510–512.

Pedley, T. J., ed. (1977). "Scale Effects in Animal Locomotion." Academic Press, Orlando, FL.

Puckett, K. J., and Dill, L. M. (1984). Cost of sustained and burst swimming to juvenile coho salmon (*Oncorhynchus kisutch*). *Can. J. Fish. Aquat. Sci.* **41,** 1546–1551.

Puckett, K. J., and Dill, L. M. (1985). The energetics of feeding territorially in juvenile coho salmon (*Oncorhynchus kisutch*). *Behaviour* **92,** 97–111.

Randall, D. J., and Brauner, C. (1991). Effects of environmental factors on exercise in fish. *J. Exp. Biol.* **160,** 113–126.

Rome, L. C., Funke, R. P., Alexander, R. McN., Lutz, G., Aldridge, H., Scott, F., and Freadman, M. (1988). Why animals have different muscle fibre types. *Nature (London)* **335,** 824–827.

Rome, L. C., Funke, R. P., and Alexander, R. McN. (1990). The influence of temperature on muscle velocity and sustained swimming performance in swimming carp. *J. Exp. Biol.* **154,** 163–178.

Rosen, D. E. (1982). Teleostean interrelationships, morphological function and evolutionary inference. *Am. Zool.* **22,** 261–273.

Sharp, G. D., and Pirages, S. W. (1978). "The Physiological Ecology of Tunas." Academic Press, Orlando, FL.

Shtaf, L. G., Pavlov, D. S., Skorobogativ, M. A., and Baryekian, A. S. (1983). Fish behaviour as affected by the degree of flow turbulence. *Vopr. Ikhtiol.* **3,** 307–317.

Skadsen, J. M., Webb, P. W., and Kostecki, P. T. (1980). Measurement of sublethal metabolic stress in rainbow trout (*Salmo gairdneri*) using automated respirometry. *J. Environ. Sci. Health, Part B* **B15,** 193–206.

Somero, G. N., and Childress, J. J. (1980). A violation of the metabolism-size scaling paradigm: Activities of glycolytic enzymes in muscle increase in larger-sized fish. *Physiol. Zool.* **53,** 322–337.

Sutterlin, A. M., and Waddy, S. (1975). Possible role of the posterior lateral line in obstacle entrainment by brook trout (*Salvelinus fontinalis*). *J. Fish. Res. Board Can.* **32,** 2441–2446.

Videler, J. J. (1981). Swimming movements, body structure and propulsion in cod *Gadus morhua*. *Symp. Zool. Soc. London* **48,** 1–27.

Videler, J. J., and Hess, F. (1984). Fast continuous swimming by two pelagic predators, saithe (*Pollachius virens*) and mackerel (*Scomber scombrus*): A kinematic analysis. *J. Exp. Biol.* **109,** 209–228.

Videler, J. J., and Wardle, C. S. (1978). New kinematic data from high speed cine film recordings of swimming cod (*Gadus morhua*). *Neth. J. Zool.* **28,** 465–484.
Videler, J. J., and Wardle, C. S. (1991). Fish swimming stride by stride: Speed limits and endurance. *Rev. Fish Biol. Fish.* **1,** 23–40.
Videler, J. J., and Weihs, D. (1982). Energetic advantages of burst-and-coast swimming of a fish at high speeds. *J. Exp. Biol.* **97,** 169–178.
Vinyard, G. L. (1982). Variable kinematics of Sacramento perch (*Archoplites interruptus*) capturing evasive and non-evasive prey. *Can. J. Fish. Aquat. Sci.* **39,** 208–211.
Vlymen, W. J. (1974). Swimming energetics of larval anchovy, *Engraulis mordax*. *Fish. Bull.* **72,** 885–899.
Wardle, C. S. (1975). Limit of fish swimming speed. *Nature (London)* **255,** 725–727.
Wardle, C. S. (1977). Effects of size on the swimming speeds of fish. *In* "Scale Effects in Animal Locomotion" (T. J. Pedley, ed.), pp. 299–313. Academic Press, Orlando, FL.
Wardle, C. S., and He, P. (1988). Burst swimming speeds of mackerel, *Scomber scombrus* L. *J. Fish Biol.* **35,** 129–137.
Wardle, C. S., and Kanwisher, J. W. (1974). The significance of heart rate in free swimming cod, *Gadus morhua*: Some observations with ultra-sonic tags. *Mar. Behav. Physiol.* **2,** 311–324.
Ware, D. M. (1975). Growth, metabolism and optimal swimming speed of a pelagic fish. *J. Fish. Res. Board Can.* **32,** 33–41.
Ware, D. M. (1978). Bioenergetics of pelagic fish: Theoretical change in swimming speed and ration with body size. *J. Fish. Res. Board Can.* **35,** 220–228.
Weatherley, A. H., Rogers, S. C., Pinock, D. G., and Patch, J. R. (1982). Oxygen consumption of active rainbow trout, *Salmo gairdneri* Richardson, derived from electromyograms obtained by radiotelemetry. *J. Fish Biol.* **20,** 479–489.
Webb, P. W. (1973). Kinematics of pectoral-fin propulsion in *Cymatogaster aggregata*. *J. Exp. Biol.* **59,** 697–710.
Webb, P. W. (1975a). Hydrodynamics and energetics of fish propulsion. *Bull. Fish. Res. Board Can.* **190,** 1–159.
Webb, P. W. (1975b). Acceleration performance of rainbow trout, *Salmo gairdneri*, and green sunfish, *Lepomis cyanellus*. *J. Exp. Biol.* **63,** 451–465.
Webb, P. W. (1976). The effect of size on the fast-start performance of rainbow trout *Salmo gairdneri* (Richardson) and a consideration of piscivorous predator–prey interactions. *J. Exp. Biol.* **65,** 157–177.
Webb, P. W. (1978a). Partitioning of energy into metabolism and growth. *In* "Ecology of Freshwater Fish Production" (S. D. Gerking, ed.), pp. 184–214. Blackwell, Cambridge, UK.
Webb, P. W. (1978b). Fast-start performance and body form in seven species of teleost fish. *J. Exp. Biol.* **74,** 211–226.
Webb, P. W. (1978c). Temperature effects on acceleration of rainbow trout (*Salmo gairdneri*). *J. Fish. Res. Board Can.* **35,** 1417–1422.
Webb, P. W. (1981). Responses of northern anchovy, *Engraulis mordax*, larvae to predation by a biting planktivore, *Amphiprion percula*. *Fish. Bull.* **79,** 727–735.
Webb, P. W. (1982a). Fast-start resistance of trout. *J. Exp. Biol.* **96,** 93–106.
Webb, P. W. (1982b). Locomotor patterns in the evolution of actinopterygian fishes. *Am. Zool.* **22,** 329–342.
Webb, P. W. (1983). Speed, acceleration and manoeuverability of two teleost fishes. *J. Exp. Biol.* **102,** 115–122.
Webb, P. W. (1984a). Body form, locomotion and foraging in aquatic vertebrates. *Am. Zool.* **24,** 107–120.

Webb, P. W. (1984b). Body and fin form and strike tactics of four teleost predators attacking fathead minnow prey. *Can. J. Fish. Aquat. Sci.* **41,** 157–165.
Webb, P. W. (1986a). Locomotion and predator–prey relationships. *In* "Predator–Prey Relationships" (M. E. Feder and G. V. Lauder, eds.), pp. 24–41. Univ. of Chicago Press, Chicago.
Webb, P. W. (1986b). Effect of body form and response threshold on the vulnerability of four species of teleost prey attacked by largemouth bass. *Can. J. Fish. Aquat. Sci.* **43,** 763–771.
Webb, P. W. (1988a). Simple physical principles and vertebrate aquatic locomotion. *Am. Zool.* **28,** 709–725.
Webb, P. W. (1988b). "Steady" swimming kinematics of tiger musky, an esociform accelerator, and rainbow trout, a generalist cruiser. *J. Exp. Biol.* **138,** 51–69.
Webb, P. W. (1989). Station holding by three species of benthic fishes. *J. Exp. Biol.* **145,** 303–320.
Webb, P. W. (1991). Composition and mechanics of routine swimming of rainbow trout, *Oncorhynchus nerka. Can. J. Fish. Aquat. Sci.* **48,** 583–590.
Webb, P. W. (1994). The biology of fish swimming. *In* "Mechanics and Physiology of Animal Swimming" (L. Maddock, Q. Bone, and J. M. V. Rayner, eds.). Harvard University Press, Cambridge, MA. In press.
Webb, P. W., and Blake, R. W. (1985). Swimming. *In* "Functional Vertebrate Morphology" (M. Hildebrand, D. M. Bramble, K. F. Liem, and D. B. Wake, eds.), pp. 110–128. Harvard Univ. Press, Cambridge, MA.
Webb, P. W., and Carolla, R. T. (1981). Burst swimming performance of northern anchovy, *Engraulis mordax*, larvae. *Fish. Bull.* **79,** 143–150.
Webb, P. W., and de Buffrénil, V. (1990). Locomotion in the biology of large aquatic vertebrates. *Trans. Am. Fish. Soc.* **119,** 629–641.
Webb, P. W., and Skadsen, J. M. (1980). Strike tactics of *Esox. Can. J. Zool.* **58,** 1462–1469.
Webb, P. W., and Smith, G. R. (1980). Function of the caudal fin in early fishes. *Copeia,* pp. 559–562.
Webb, P. W., and Weihs, D. (1986). Functional locomotor morphology of early life history stages of fishes. *Trans. Am. Fish. Soc.* **115,** 115–127.
Webb, P. W., Kostecki, P. T., and Stevens, E. D. (1984). The effect of size and swimming speed on locomotor kinematics of rainbow trout. *J. Exp. Biol.* **109,** 77–95.
Webb, P. W., Sims, D., and Schultz, W. W. (1991). The effect of an air/water interface on the fast-start performance of rainbow trout (*Oncorhynchus mykiss*). *J. Exp. Biol.* **155,** 219–226.
Webb, P. W., Hardy, D. G., and Mehl, V. L. (1992). The effect of armored skin on the swimming of longnose gar, *Lepisosteus osseus. Can. J. Zool.* **70,** 1173–1179.
Weihs, D. (1972). A hydrodynamic analysis of fish turning manoeuvers. *Proc. R. Soc. London, Ser. B* **182,** 59–72.
Weihs, D. (1973a). Mechanically efficient swimming techniques for fish with negative buoyancy. *J. Mar. Res.* **31,** 194–209.
Weihs, D. (1973b). The mechanism of rapid starting of slender fish. *Biorheology* **10,** 343–350.
Weihs, D. (1973c). Optimal fish cruising speed. *Nature (London)* **245,** 48–50.
Weihs, D. (1974). The energetic advantages of burst swimming. *J. Theor. Biol.* **49,** 215–229.
Weihs, D. (1975). An optimum swimming speed of fish based on feeding efficiency. *Isr. J. Technol.* **13,** 163–167.

Weihs, D. (1980). Energetic significance of changes in swimming modes during growth of anchovy larvae, *Engraulis mordax*. *Fish. Bull.* **77,** 597–604.

Weihs, D. (1981). Effects of swimming path curvature on the energetics of fish motion. *Fish. Bull.* **79,** 171–176.

Weihs, D., and Webb, P. W. (1983). Optimization of locomotion. *In* "Fish Biomechanics" (P. W. Webb and D. Weihs, eds.), pp. 339–371. Praeger, New York.

Werner, E. E. (1986). Species interactions in freshwater fish communities. *In* "Community Ecology" (J. Diamond and T. J. Case, eds.), pp. 334–358. Harper & Row, New York.

Wood, C. M. (1991). Acid–base and ion balance, metabolism, and their interactions, after exhaustive exercise in fish. *J. Exp. Biol.* **160,** 285–308.

Wood, C. M., Turner, J. D., and Graham, M. S. (1983). Why do fish die after severe exercise? *J. Fish Biol.* **22,** 189–201.

Wu, T. Y., Brokaw, C. J., and Brennen, C. (eds) (1975). "Swimming and Flying in Nature." Plenum Press, New York, New York.

Jumping Ability of Anuran Amphibians

RICHARD L. MARSH

Department of Biology, Northeastern University, Boston, Massachusetts 02115

I. Introduction
II. Mechanics of Jumping
  A. Derivation of the Formulae for Jumping Distance
  B. Influence of Takeoff Angle and Ground Reaction Force on Jumping
  C. Calculation of Mechanical Work and Power from Jumping Distance
III. Providing the Power: The Muscles Used in Jumping
  A. Contractile Properties
  B. Muscles Used in Jumping and Their Dimensions
IV. Jumping Performance
  A. Interspecific Variation in Jumping Performance Including Allometry
  B. Intraspecific Allometry
  C. Predicted Allometric Scaling
  D. Work and Power Output
  E. Thermal Effects
V. Detailed Studies of Jumping
  A. Ground Reaction Forces
  B. Kinematics
  C. Muscle Activity and Shortening
VI. Conclusions
VII. Abbreviations
References

I. Introduction

The goal of this review is limited to an analysis of the jumping ability of anuran amphibians. The reasons for this emphasis on a single facet of amphibian exercise performance are several. First, the specialization for jumping is such an obvious aspect of the biology of anurans. Most anurans employ at least partly a saltatory gait referred to as hopping, leaping, or jumping. Some species have become secondarily specialized for swimming, walking, or burrowing, but all bear the imprint of ancestors that were jumpers (Zug, 1978). Second, the jumping performance of frogs provides a focus for integrating several types of studies of exercise performance. Frog muscle studied *in vitro* has formed the basis for much of our knowledge of the mechanics and energetics of skeletal muscle (Woledge *et al.*, 1985). However, few studies have attempted to integrate this knowledge with investigations of *in vivo* performance. Authors that have considered this integration have reached widely differing conclusions regarding the relation between *in vivo* and *in vitro* muscle performance. For example, Hirano and Rome (1984) concluded that jumping performance is a predictable consequence of the intrinsic properties of the muscles that power the jump. Conversely, Emerson (1985) suggested that the distance jumped by small frogs greatly exceeds that predictable from the performance of isolated muscles. Finally, many other aspects of amphibian performance have been very thoroughly reviewed, including aerobic and anaerobic performance during sustainable and nonsustainable exercise (Gatten *et al.*, 1992) and performance during calling and other natural activities (Pough *et al.*, 1992). Little of substance can be added to these accounts.

This review will analyze both the basic biomechanics of the jump and the intrinsic properties of the muscles that power jumping in anurans. Together these analyses ought to be able to predict jumping performance and can be used to interpret past studies of jumping in anurans and provide a firmer basis for future studies. Additionally, an analysis of mechanical performance of anuran muscles during jumping provides provides the background to evaluate the *in vitro* performance of anuran muscles.

II. Mechanics of Jumping

A. Derivation of the Formulae for Jumping Distance

Jumping is a relatively simple movement when analyzed in terms of the dynamics of the center of mass of the animal. However, previous authors have sometimes ignored certain characteristics of the jump in an attempt to simplify the predictive equations. Oversimplification may lead to erroneous conclusions regarding aspects of performance such as the optimum predicted jumping angle, allometry, and thermal effects. Many of the basic equations that govern jumping performance are derived in introductory physics textbooks (e.g., Beiser, 1978) and have been used to analyze movements in animals and humans (e.g., Kreighbaum and Barthels, 1990). A particularly useful analysis of jumping height is provided by Bennet-Clark (1977), and many of the same principles apply to jumping distance.

The horizontal distance jumped by a frog can be subdivided into several components (Fig. 1). The level distance traveled by the center of mass between the time of takeoff and the time it returns to the takeoff height ($L_{j,l}$) has been discussed most often (e.g., Emerson, 1985). This distance can be derived from very simple ballistics. If lift and drag forces are small (Bennet-Clark, 1977), then the horizontal velocity will be constant after the frog leaves the ground. Thus,

$$L_{j,l} = V_h t_{a,l}, \tag{1}$$

where V_h is the horizontal velocity and $t_{a,l}$ is the time spent in the air until the center of mass returns to the takeoff height (H_t). The increase in height after takeoff (H_a) is calculated as

$$H_a = \frac{V_v^2}{2g}, \tag{2}$$

where V_v is the vertical component of velocity and g is the acceleration due to gravity. The time spent in the air until the center of mass returns to the takeoff height is then derived as

$$t_{a,1} = 2\left(\frac{2H_a}{g}\right)^{0.5} \quad \text{or} \quad t_{a,1} = \frac{2V_v}{g}. \tag{3}$$

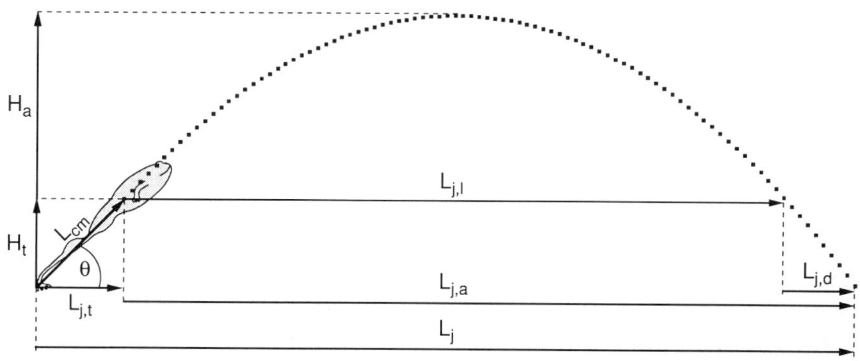

FIG. 1. Idealized jump by an anuran amphibian. The dotted line shows the trajectory of the center of mass based on simple ballistics (ignoring lift and drag). The approximately 1-m jump diagrammed is based on a 65-g frog jumping with a takeoff velocity of 2.8 m s^{-1} (V_h = 2 m s^{-1}) and a takeoff angle of 45°. (The optimum angle for this jump would actually be 41°.)

The vertical and horizontal velocities can be derived by resolving the takeoff velocity (V_t) as

$$V_v = V_t \sin \theta \tag{4}$$

$$V_h = V_t \cos \theta, \tag{5}$$

where θ is the takeoff angle. From Eqs. (1), (2), and (3) and the trigonometric identity $2 \sin \theta \cos \theta = \sin 2\theta$, jumping distance is calculated as

$$L_{j,l} = \frac{V_t^2 \sin 2\theta^*}{g}. \tag{6}$$

This formula leads to the often cited prediction that the optimum takeoff angle for maximal jumping distance is 45°.

However, the center of mass at takeoff is considerably above that at landing because of the long hindlimbs of the frog (Gray, 1968). Thus, the time taken to rise to the maximum height after takeoff (t_r) is shorter than the time required to fall to the ground (t_f):

$$t_r = \left(\frac{2H_a}{g}\right)^{0.5} \tag{7}$$

$$t_f = \left(\frac{2(H_a + H_t)}{g}\right)^{0.5}. \tag{8}$$

Substituting for H_a results in

$$t_r = \frac{V_t \sin \theta}{g} \tag{9}$$

$$t_f = \frac{(V_t^2 \sin^2 \theta + 2gH_t)^{0.5}}{g}. \tag{10}$$

These relations result in a rather unfortunately complicated formula for aerial jumping distance:

$$L_{j,a} = \frac{V_t^2 \sin \theta \cos \theta + V_t \cos \theta (V_t^2 \sin^2 \theta + 2gH_t)^{0.5}}{g}. \tag{11}$$

One important result of this relation is that there is no longer one optimum angle for maximum jumping distance; instead, the optimum angle depends on V_t and H_t. This dependence has been emphasized in discussing the performance of humans in jumping and throwing heavy weights (Kreighbaum and Barthels, 1990). For anuran jumping the optimum angle is directly related to the force exerted on the ground expressed as a multiple of body weight (see Fig. 3B). This dependence means that the optimum angle is lower for shorter jumps and larger frogs. The importance of takeoff height was recognized by Calow and Alexander (1973), who calculated that their frogs (*Rana temporaria* weighing approximately 25 g) should have jumped at an angle of approximately 34° to obtain maximum distance. However, their derivation assumed incorrectly that H_t and V_t were constants independent of takeoff angle.

The center of mass is initially located approximately above the hindfeet (Hirsch, 1931). The forward and upward movements of the center of mass depend on extension of the hindlimbs and thus are functions of the takeoff angle (Fig. 1). I will assume here that the body and limbs are aligned at takeoff and that the angle between the long axis of the limbs and the ground is the same as the takeoff angle. [This assumption appears correct for at least some small frogs, including ranids (Hirsch, 1931; Calow and Alexander, 1973) and hylids (Marsh and John-Alder, 1994, see Fig. 14), but may be incorrect for large frogs such as the American bullfrog *Rana catesbeiana* (R. L. Marsh and J. M. Olson, unpublished), in which the urostyle may begin to flex before takeoff.] Given this assumption, the vertical and horizontal distances moved by the center of mass before takeoff are calculated as

$$H_t = L_{cm} \sin \theta \tag{12}$$

$$L_{j,t} = L_{cm} \cos \theta, \tag{13}$$

where L_{cm} is the distance from the center of mass to the tip of the toes along the extended hindlimbs (Fig. 1). Total jumping distance is then

$$L_j = \frac{V_t^2 \sin \theta \cos \theta + V_t \cos \theta (V_t^2 \sin^2 \theta + 2gL_{cm} \sin \theta)^{0.5}}{g} + L_{cm} \cos \theta. \tag{14}$$

This formula provides an accurate estimate of jumping distance (Table I).

One final complication in the relation between takeoff angle and distance is the influence of angle on takeoff velocity. If the hindlimbs are capable of producing a certain maximum force independent of angle, then takeoff velocity will decrease with increasing angle owing to the opposing force of gravity (Fig. 2A) (Tricker and Tricker, 1967). This prediction is, however, not entirely straightforward because the overall acceleration results from a force pulse produced by the hindlimb muscles. The more rapidly these muscles shorten, the less force they will produce (see Fig. 5).

TABLE I

COMPARISON OF ACTUAL AND PREDICTED JUMPING DISTANCES IN *RANA CATESBEIANA*[a]

	M_b (g)	L_{sv} (m)	θ (°)	V_t (m s^{-1})	L_j (m) Actual	Predicted	Predicted/Actual
	534	0.172	28	2.39	0.98	1.02	1.05
			41	2.24	1.03	0.96	0.94
			31	2.49	0.90	1.06	1.18
	470	0.158	31	2.91	1.08	1.32	1.22
			25	2.71	1.13	1.18	1.05
			37	2.50	0.93	1.07	1.16
	523	0.162	55	2.56	1.08	0.89	0.83
			46	2.35	1.10	0.93	0.85
			50	2.52	1.10	0.96	0.88
	356	0.162	43	2.61	1.00	1.08	1.08
			34	2.49	1.20	1.06	0.88
			36	2.55	1.00	1.05	1.05
Means	471	0.164	42	2.53	1.04	1.05	1.01

[a] Frogs were jumped with a T_b of approximately 20°C. Take off angle (θ) and velocity (V_t) and actual jumping distance were measured from video recordings. Predicted jumping distance was calculated from Eq. (14). From R. L. Marsh and J. M. Olson (unpublished).

B. Influence of Takeoff Angle and Ground Reaction Force on Jumping

To illustrate selected characteristics of jumping I have calculated the jumping performance of a 0.03-kg frog using the relations described here. This size frog was chosen because it corresponds to that used in several past studies of frog jumping (Hirsch, 1931; Calow and Alexander, 1973; Hirano and Rome, 1984; see also Fig. 15). First, the influence of changing the angle at which force is exerted was examined while maintaining the average ground reaction force constant at 3.5 times the body weight (Fig. 2). This force produces a total jumping distance of 0.85 m, which is close to the mean maximum jumping distance predicted for this size frog at 23°C (Section IV, A). The optimum angle for maximum total jumping distance is calculated to be 39° (Fig. 2B). The angle that produces the maximum $L_{j,1}$ is higher, 42°, but is still lower than the 45° predicted by Eq. (6), because velocity drops as the angle at which the force is applied increases (Fig. 2A). At the optimum angle for maximum jumping distance, the total distance traveled is 21% higher than that predicted by Eq. (6) because of additional distances traveled before takeoff ($L_{j,t}$) and after the frog descends below takeoff height ($L_{j,d}$). The total work done during takeoff changes only slightly with increasing angle (Fig. 2C), because the decreasing kinetic energy due to lower V_t is offset by the increasing potential energy of the center of mass. The total work required to move a unit distance is thus also minimized at the angle that maximizes jumping distance. Noteworthy is the fact that the frog can take off over a broad range of jumping angles and still achieve nearly maximal

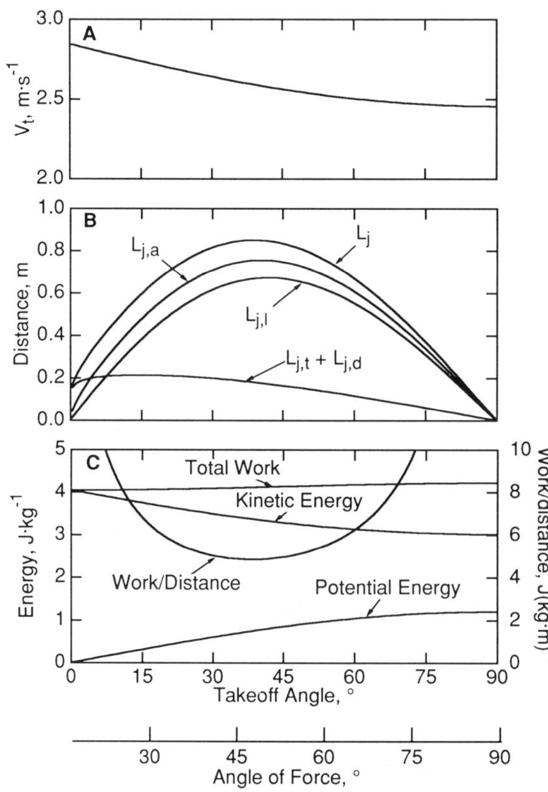

FIG. 2. The influence of takeoff angle (θ) on selected characteristics of jumping. The data were calculated for a 0.03-kg frog (L_{cm} = 0.12 m) exerting an average ground reaction force of 1 N (3.5 M_b g) during takeoff. This force was applied at varying angles and V_v, V_h, and θ were calculated from the respective vertical and horizontal forces. (A) Takeoff velocity calculated from V_v and V_h. (B) Jumping distances calculated from V_t, θ, and L_{cm} using Eqs. (6) and (11–14). (C) Body-mass-specific work and work per unit jumping distance.

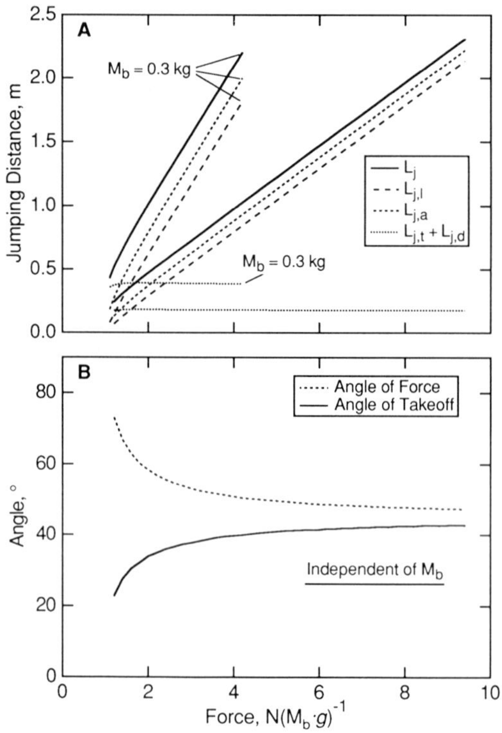

FIG. 3. The influence of ground reaction force on the characteristics of jumping. Unless indicated the values are for a 0.03-kg frog (L_{cm} = 0.12 m). Selected values are plotted for a 0.3-kg frog (L_{cm} = 0.26 m). The values shown resulted from applying the force at varying angles and reporting the values at the angle that resulted in maximal total jumping distance. (A) Jumping distances for 0.03- and 0.3-kg frogs. (B) Optimum angle for the application of the ground reaction force and optimum takeoff angle. These angles are independent of body mass. (C) Distances traveled during takeoff ($L_{j,t}$) and after returning to takeoff height ($L_{j,d}$) for a 0.03-kg frog. (D) Takeoff velocities and contact times for a 0.03-kg frog. (E) Energy at takeoff and energy per unit jumping distance for 0.03- and 0.3-kg frogs. The energy per unit jumping distance is independent of body mass.

distance. Total jumping distance drops to 90% of its maximal value when angle is decreased to 25° or increased to 54° (see also Gray, 1968).

The influence of changing the mean ground reaction force produced by the hindlimbs is illustrated in Fig. 3. These results were obtained by calculating jumping performances at varying angles of force application and reporting the values at the angle that resulted in maximum total jumping distance. Performance is calculated for a 0.03-kg frog jumping with increasing mean ground reaction forces to a maximum of six times body weight, which predicts a jump in excess of

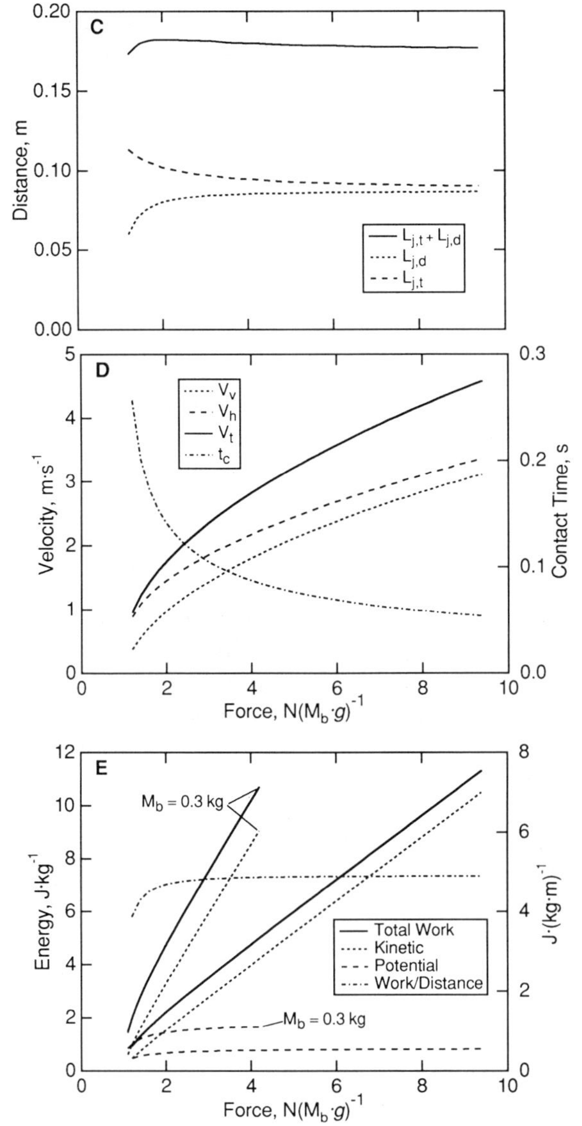

FIG. 3. *(Continued)*

2 m (Fig. 3A). This jumping distance is near the longest documented jump for this size frog (Zug, 1978). The optimum takeoff angle increases as the applied force increases, whereas the angle at which the force must be exerted decreases (Fig. 3B). The angle at which the force is applied is always higher than the takeoff angle because the force

required to support the body weight is a significant fraction of the vertical force. Total jumping distance increases linearly with the force applied. This increase is almost entirely due to an increase in $L_{j,l}$, and an approximately constant increment over this distance is provided by the sum of $L_{j,t}$ and $L_{j,d}$. This constant increment is due to offsetting decreases and increases in $L_{j,t}$ and $L_{j,d}$, respectively (Fig. 3C). Predicted contact time decreases with increasing force, corresponding to the increase in average velocity and thus takeoff velocity (Fig. 3D). The amount of energy expended increases linearly with the force applied, and the energy required to move a unit distance is relatively constant independent of jumping distance (Fig. 3E). (These calculations are probably not entirely realistic for very low forces because they assume that the frog fully extends its hindlimbs regardless of force, and this is likely not to be the case (J. M. Olson and R. L. Marsh, unpublished).

C. Calculation of Mechanical Work and Power from Jumping Distance

The literature on anuran jumping performance contains mostly information on distance only. The preceding analysis provides the basis for calculating the mechanical work and power output required to produce the measured jumping distances. A necessary assumption is that the jumps are undertaken at an angle near the optimum. This assumption is probably reasonable when investigators report maximal jumping distances (Section IV). The relative contribution of moving the center of mass upward and forward during takeoff to total jumping distance varies with the distance jumped and the size of the frog (Fig. 3A). If a particular frog jumps at or near the optimum takeoff angle, the contribution of $(L_{j,t} + L_{j,d})$ to jumping distance is relatively constant (Fig. 3C). This distance can be approximated by the value $2L_{cm} \sin 45°$. No precise measures of L_{cm} have been reported. However, based on the position of the center of mass at takeoff estimated by Hirsch (1931) on *Rana ridibunda*, and estimates made on *Osteopilus* and *Rana catesbeiana* (R. L. Marsh, unpublished), this value appears to be approximately 1.1–1.2 times the length of the outstretched hindlimbs (L_{hl}). The relative length of the hindlimbs varies in different anurans. This length has often been expressed as a fraction of the more commonly measured snout–vent length (L_{sv}). The ratio L_{hl}/L_{sv} varies from about 1.25 for anurans classed as relatively poor jumpers, for example, bufonid toads, to greater than 1.9 for species with the best jumping ability, for example, the hylid *Acris gryllus*. Unfortunately, the L_{hl}/L_{sv} ratios reported in the literature vary with the method used to estimate L_{hl}. Some investigators report L_{hl} as the sum of individual segment lengths (Dobrowolska, 1973; Emerson, 1978). This method appears to overestimate the L_{hl} measured directly on the outstretched hindlimb (Zug and Altig, 1978).

Estimating $L_{j,t} + L_{j,d}$ allows the calculation of $L_{j,l}$ from the total jumping distance. Calculating $L_{j,l}$ permits takeoff velocity to be estimated from Eq. (6) if takeoff angle is known. Actually, takeoff angle need not be known with any great degree of precision, as using any angle from 30° to 45° results in only a few percent error in calculating takeoff velocity. Accurate prediction of V_t allows the contact time to be calculated using Eqs. (16) and (17).

The total work expended in the jump is, of course, dependent on the sum of the kinetic and potential energies at takeoff. Most past considerations of jumping have considered only the kinetic energy, but the contribution of the potential energy can be substantial, particularly in submaximal jumps and for large frogs (Fig. 3E). However, because the expenditure of energy to raise the center of mass contributes to the the total jumping length, using total jumping length to calculate total work using Eq. (6) overestimates kinetic energy, but results in a close approximation of the the total work. In fact, agreement is better if a constant angle of 45° is assumed, rather than the actual takeoff angle.

Jumping performance is determined by the production of mechanical power by the skeletal muscles in the hindlimb. Therefore, the relation between power requirements and jumping distance assumes central importance in discussing the jumping performance of frogs. The power expended in jumping can be derived in a straight-forward manner if one assumes the simplest ballistics model as expressed in Eq. (6) (see also Bennet-Clark, 1977). Average power expended to accelerate the body mass during takeoff ($\dot{W}_{j,k}$) is simply the kinetic energy present in the body mass (M_b) at takeoff divided by the time required (t_c):

$$\dot{W}_{j,k} = \frac{M_b V_t^2}{2t_c}. \tag{15}$$

The contact time can in turn be calculated as

$$t_c = \frac{V_t M_b}{F_t}, \tag{16}$$

where F_t is the average force applied during takeoff (not including the force required to lift the body weight). The force in turn can be reduced to

$$F_t = \frac{V_t^2 M_b}{2L_{cm}}. \tag{17}$$

TABLE II

JUMPING PERFORMANCE OF FIVE SPECIES OF HYLID TREE FROGS[a]

Species[b]	n[c]	Mass (g)	L_{hl} (mm)	t_c (ms) Actual	t_c (ms) Predicted[d]	V_t (m s^{-1})	W_k (J kg^{-1})	W_p (J kg^{-1})	W_{tot} (J kg^{-1})	$\dot{W}_j$ (W kg^{-1})
O.s.	7(2)	27.9	110	98	102	2.52	3.17	0.72	3.93	40.9
H.c.	1(1)	8.7	70	75	81	1.99	1.98	0.59	2.57	34.2
H.s.	2(2)	3.8	55	63	68	1.83	1.68	0.36	2.04	33.2
P.c.	6(2)	0.80	36	39	37	2.27	2.60	0.29	2.88	76.9
A.g.	1(1)	0.63	34	35	34	2.26	2.55	0.29	2.84	80.5

[a] Values reported are for the maximal jumps recorded. For *O.s.* and *P.c.* several nearly identical jumps were obtained on each animal and mean values for all jumps are presented. Frogs were filmed at 303 Hz and data were obtained from frame-by-frame analysis. Data presented in a slightly different form in Marsh and John-Adler (1994). See Section VII for abbreviations.

[b] Species tested: *O.S.*, *Osteopilus septendrionalis; H.c.*, *Hyla cinerea; H.s.*, *Hyla squirella; P.c.*, *Pseudacris crucifer; A.g.*, *Acris gryllus.*

[c] Number of jumps (number of animals).

[d] Predicted t_c from Eq. (16) and (17).

(Bennet-Clark, 1977). Combining Eqs. (16) and (17) results in a realistic prediction of contact time at least for small hylid frogs (Table II).

From the preceding and Eq. (6), jumping distance can be derived as a function of power, body mass, and the distance moved during takeoff:

$$L_{j,1} = \left(\frac{\dot{W}_{j,k} L_{cm}}{M_b}\right)^{2/3} \cdot \frac{\sin 2\theta}{g}. \quad (18)$$

[This equation for jumping distance is similar to one of Bennet-Clark (1977) for jumping height, except his equation uses peak power, which in his model is twice the average power.] Equation (18) does not include the power required to lift the body weight to H_t. If this equation is rearranged to calculate power from jumping distance, it will underestimate the total average power ($\dot{W}_j$) if $L_{j,1}$ is used, but will overestimate $\dot{W}_j$ if the total jump distance is used. A close approximation of $\dot{W}_j$ is obtained by using $L_{j,a}$ instead (see also Hirano and Rome, 1984). Alternatively, work can be obtained as indicated in the foregoing and power calculated by dividing this total work by the contact time derived from Eqs. (16) and (17).

III. Providing the Power: The Muscles Used in Jumping

A. Contractile Properties

The contractile properties of anuran hindlimb muscles, particularly those from several species of *Rana*, have probably been studied more than those of any other skeletal muscles (Hill, 1965, 1970; Woledge *et al.*, 1985). Much of this vast literature is oriented toward understanding the underlying basis of muscular contraction, and not specifically toward applying this knowledge to *in vivo* performance. Two reviews related specifically to amphibian skeletal muscle have attempted to shift this focus somewhat (Gans and de Gueldre, 1992; Rome *et al.*, 1992). General features of vertebrate skeletal muscle relevant to exercise performance are also reviewed elsewhere in this volume (see paper by Rome in Volume 38A). Nevertheless, interpreting many of the studies of frog muscle in terms that are relevant to jumping performance is difficult without further knowledge of just how the muscles function during jumping (Section IV, C). Direct application of many *in vitro* studies to jumping is also hindered because the muscles were studied at 0°C, a temperature at which most frogs do not jump well. I will concentrate here on the *in vitro* properties of frog muscle that should broadly correlate with the ability of anuran muscle to produce mechanical work and power at temperatures similar to those used in studies of jumping performance.

Before examining the relations between contractile properties and jumping performance, some consideration must be given to the diversity of fiber types found in amphibians. Gans and de Gueldre (1992) have summarized much of what is known regarding amphibian fiber types. The best jumpers studied all have predominantly amphibian Type 1 fibers (fast, glycolytic) in their hindlimb muscles (Sperry, 1981; Putnam and Bennett, 1983; Marsh and Taigen, 1987; Mendiola *et al.*, 1991). This fiber type occupies between 73 and 90% of the cross-sectional area in the major hindlimb muscles of *Rana* and *Hyla*. The remaining cross-sectional area is occupied almost entirely by Type 2 (fast, oxidative, glycolytic) fibers. The situation is somewhat less clear in toads of the genus *Bufo*. Putnam and Bennett (1983) report that in *Bufo boreas* the distribution of fiber types is similar to that found in *Rana* and the aquatic *Xenopus*. Their data suggest that the major difference among these species is in the quantitative distribution of aerobic enzymes. However, Mendiola *et al.* (1991) and Sperry (1981) report that bufonids have predominantly amphibian Type 2 fibers in their hindlimb muscles. Data on contractile properties of the sartorius from *Bufo americanus* are compatible with the view that this muscle contains mostly Type 2 fibers (Table III and Fig. 6), but in the absence of data on isolated fibers from *Bufo* any conclusion must be considered tentative. Most of the hindlimb muscles, including those muscles on which contractile properties commonly have been measured, have a similar distribution of fiber types (Sperry, 1981; Putnam and Bennett, 1983; Mendiola *et al.*, 1991). This consistency in fiber type composition makes correlations between contractile properties of individual muscles and jumping performance more robust.

Isometric contractile data provide information related to muscle strength and the rapidity of activation and deactivation. Intrinsic strength is usually referenced to the maximal active force (P_0) in a fully fused isometric tetanus at optimal length (L_0), and expressed as force per cross-sectional area (stress). Many conditions pertaining *in vivo* may alter the actual force produced during jumping, including starting length, distance shortened, length change prior to the jump, and velocity of shortening during the jump (Hill, 1970; Woledge *et al.*, 1985). Nevertheless, the P_0 values provide a useful benchmark in comparing different muscles.

The intrinsic strength of amphibian twitch muscle, as measured in whole parallel-fibered muscles and single fibers, ranges from 179 to 372 kN m^{-2} with a mean of 267 $\pm$ 8 kN m^{-2} across all studies summa-

TABLE III

In Vitro Contractile Properties of Anuran Muscles[a]

M_b[b] (g)	Muscle[c]	Fiber type	T_m (°C)	P_0 (kN m^{-2})	V_{max} (L_0 s^{-1})	$R_{\dot{W}}$	$\dot{W}_{max}$ (W kg^{-1})	tP_{tW} (ms)	$t50\%R$ (ms)	Reference
Rana temporaria										
(25)	Srt.	1	0.0		1.29	0.097				Hill (1938)
			8.9		2.62	0.102				
			13.5		3.22	0.103				
			19.8		6.05	0.079				
(25)	Srt.	1	0.0		1.31	0.098				Katz (1939)
			10.9		2.94	0.118				
21–28	P.L.	1	15	350						Calow and Alexander (1973)
	Cru.	1	15	150						
	Smt.	1	15	250						
	Gr.M	1	15	340						
	Smm.	1	15	320						
(25)	Smt. (SF)	1	2.0		1.91	0.118				Edman *et al.* (1976)
(25)	Smt. (SF)	1	2.2		1.95					Edman and Hwang (1977)
(25)	T.A. & Smt. (SF)	1	2.2	220						Edman (1979)
			12.0	270						
(25)	T.A. (SF)	1 & 2(?)	15.5					47	57	Iaizzo (1990)
		1 & 2(?)	20.5					30		
	Lum. IV (SF)	1	15.5					59	24	
Rana pipiens										
(25)	Srt.	1	0.0	163	1.76	0.101	29			Fenn and Marsh (1935)
			22.0	283	8.73	0.079	196			
(25)	Srt.	1	5.0	234	1.94	0.119	54			Rome (1983)
			15.0	261	3.62	0.146	138			

(continued)

TABLE III (*Continued*)

M_b[b] (g)	Muscle[c]	Fiber type	T_m (°C)	P_0 (kN m^{-2})	V_{max} (L_0 s^{-1})	$R_{\dot{W}}$	$\dot{W}_{max}$ (W kg^{-1})	tP_{tW} (ms)	$t50\%R$ (ms)	Reference
			25.0	269	6.36	0.147	251			
38	Srt.	1	5.0	177				125		Renaud and Stevens (1981a)[d]
			15.0	260				59		
			25.0	240				23		
15–31	Srt.	1	5.0	195	2.04	0.121	47			Renaud and Stevens (1984)[d]
			15.0	240	5.26	0.134	170			
			20.0	249	7.33	0.143	260			
			25.0	249	9.76	0.129	314			
(25)	Srt.	1	0.0	253						Burchfield and Rall (1986)
			10.0	362						
Rana esculenta										
(25)	Smt. (SF)	1	19.8	291	8.60	0.137	343			Cecchi *et al.* (1978)
			4.0	213	2.37	0.134	67			
Rana catesbeiana										
15	Srt.	1	20.0	273	8.34	0.103	235	24	29	R. L. Marsh (unpublished)
375	Srt.	1	20.0	302	6.28	0.105	199	35	25	
Hyla versicolor										
7	Srt.	1	20.0	261	9.25	0.085	205	27	28	R. L. Marsh (unpublished)
Litoria aurea										
(35)	Iliof. (MU)	1	24.0					23		Luff and Proske (1979)
		1	24.0					34		
		1 & 2[e]	24.0					26		
(35)	Srt.	1, 2, & 3[e]	23.0					27		Luff and Proske (1976)
(35)	Srt.	1	10.0	233						Gibbs and Chapman (1974)
			20.0	266						

Osteopilus septentrionalis										
21	Srt.	1	20.0	317	7.98	0.089	225	31	32	R. L. Marsh (unpublished)
Bufo americanus										
21–80	Srt.	2(?)	5.0	208	1.91	0.088	35			Renaud and Stevens (1984)[c]
			10.0	236	3.07	0.096	70			
			15.0	274	4.57	0.097	121			
			20.0	274	6.10	0.102	170			
			25.0	284	7.88	0.106	237			
Bufo marinus										
130	Srt.	2(?)	10.0	252						Gibbs and Chapman (1974)
			20.0	306						
Xenopus laevis										
			10.0	275	4.35	0.129	154			
(80)	Iliof.	1	15.0	327	5.76	0.137	258			
			20.0	372	7.24	0.136	365	26	42	Lannergren *et al.* (1982)
	Iliof.	2	10	272	3.01	0.110	90			
			15	298	4.57	0.119	161			
			20.0	324	6.27	0.112	228	33	36	

[a] See Section VII for abbreviations not given in the footnotes.

[b] Body masses in parentheses are estimates based on typical masses for the species.

[c] Cru., crualis; Gr.M, gracilis major; Iliof., Iliofibularis; Lum.IV, lumbricus IV; MU, motor units; P.L., *plantaris longus* (gastrochemius); SF, single fibers; Smm., semimembranosus; Smt., semitendinosus; Srt., sartorius; T.A., tibialis anterior.

[d] Values are weighted means of data from several experimental groups.

[e] Values are weighted means based on the $\%P_0$ occupied by the motor unit.

rized in Table III. Temperature has a significant effect on P_0 in anuran skeletal muscle as in other striated muscles (Bennett, 1984). This thermal dependence is best expressed as an R_{10}, because force is not a rate function (Bennett, 1984). The influence of temperature changes substantially from 0° to 30°C (Fig. 4). This changing thermal dependence is reflected in the thermal effects on jumping distance (Section IV,E). The P_0 of anuran muscles appears to be independent of body size (Tables III and IV).

The kinetics of the isometric twitch provide an indirect assessment of events related to contractile activation and deactivation (Marsh, 1990). The rate of deactivation may limit performance during locomotion involving cyclical limb movements such as running and swimming (Marsh and Bennett, 1985, 1986b; Marsh, 1988, 1990; Else and Bennett, 1987; John-Adler *et al.*, 1989). However, jumping performance may not be closely linked to the kinetics of deactivation. In jumping, a frog spends considerable time off the ground following takeoff, during which deactivation can occur prior to repositioning the limbs for landing. For example, a 60-g frog spends about 120 ms in

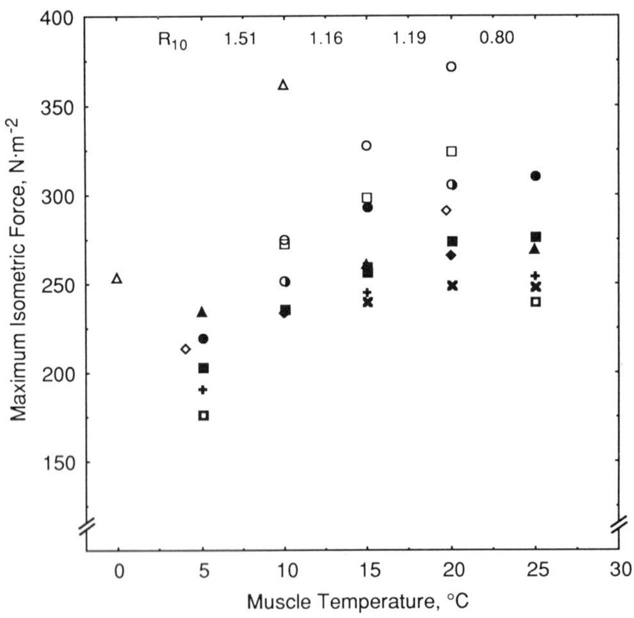

FIG. 4. Maximal isometric force in tetanic contractions as a function of muscle temperature in anuran muscles. (Different symbols represent data from the various studies reported in Table III.)

TABLE IV

ALLOMETRIC RELATIONSHIPS ($y = aM_b{}^b$; M_b in g) OF MORPHOLOGICAL AND CONTRACTILE PROPERTIES IN *RANA CATESBEIANA*[a]

y	r	a	b	95% confidence interval
Morphological Measures				
L_{sv}, mm	0.991	25.3	0.297	0.274–0.319
L_{thigh}, mm	0.997	13.2	0.287	0.271–0.304
$L_{lower\ leg}$, mm	0.996	11.2	0.305	0.284–0.326
Contractile Properties				
L_0, mm	0.992	11.0	0.293	0.272–0.313
tP_{tw}, ms	0.816	17.2	0.12	0.073–0.166
$R_{\dot{W}}$	0.294	NS[1]		
P_0, kN m^{-2}	0.29	NS		
P_{tw}, kN m^{-2}	0.044	NS		
V_{max}, L_0 s^{-1}	0.816	10.9	−0.094	(−0.134–−0.054)
$\dot{W}_{max}$, W kg^{-1}	0.318	NS		

[a] From R. L. Marsh (unpublished).
[b] NS = Not significant, $P > 0.05$.

taking off for a 1-m jump, and then spends 450 ms in the air. On the other hand, a fast rate of activation that allowed for a rapid buildup of force would presumably be advantageous. The twitch times recorded from the Type 1 fibers of 20- to 30-g frogs are approximately 25–35 ms at 20°C (Table III). [Longer twitch times have been found in studies that used transducers designed to operate with polygraph recorders, e.g., Osgood and Brewster (1963).] Twitch times are considerably shorter than the approximately 100 ms expected for contact time during jumping in this size frog at 20°C. The lack of coupling between twitch kinetics and jumping kinetics is illustrated by examining the effects of body size and temperature. The twitch becomes longer during growth in *Rana catesbeiana* (Table IV), but the allometric scaling ($M_b^{0.12}$) is much less than that expected for contact time during the jump ($M_b^{0.2}$). Also, jumping performance does not appear to be dramatically affected by the very long twitch times that occur at low temperatures (John-Alder *et al.*, 1989).

Isotonic contractile data provide information about the intrinsic shortening velocity of the contractile elements under various loads, that is, the force–velocity curve (Fig. 5). Although the muscles clearly will not be operating isotonically *in vivo* (Marsh *et al.*, 1992), these data nevertheless provide a starting point for evaluating the ability to pro-

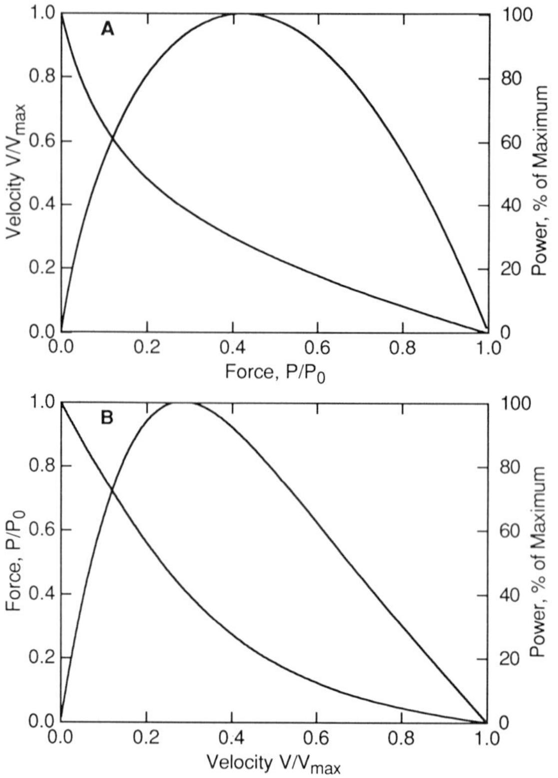

FIG. 5. Idealized isotonic force–velocity curves. The force–velocity curves have been drawn assuming the shape found for the sartorius muscle of *Rana catesbeiana*. (A) Velocity and power versus force. (B) Force and power versus velocity.

duce mechanical power. For our purposes the important characteristics of the curve can be summarized by the maximum velocity of shortening at zero load as extrapolated from the force–velocity curve (V_{max}) and the the power ratio, $R_{\dot{W}} = \dot{W}_{max}(V_{max}P_0)^{-1}$, where $\dot{W}_{max}$ is the maximum power (Marsh and Bennett, 1985).

Not surprisingly given their fiber type composition, the isotonic properties of whole ranid and hylid muscles closely resemble those of single Type 1 fibers or motor units (Table III). Recent studies using servocontrolled ergometers of six species of small (<50 g) frogs show consistent values averaging $8.1L_0$ s^{-1} for V_{max} at 20°C (Table III). Several studies of the sartorius muscle from *Rana temporaria* and *Rana pipiens* using mechanical lever systems have found substantially lower V_{max} values over a range of muscle temperatures (Hill, 1938;

Katz, 1939; Rome, 1983; but see Fenn and Marsh, 1935). Allowing for the effects of temperatures, the muscles of anurans have to be considered among the fastest muscles measured to date. The fastest V_{max} values reported in the literature are approximately $24L_0$ s^{-1} in fast mouse muscles measured at 35°C (Close, 1965; Luff, 1981), which given a Q_{10} of 2, corresponds to about $8.5L_0$ s^{-1} at 20°C. Muscles from the toad *Bufo americanus*, a relatively poor jumper (Section IV, A), and the Type 2 fibers from *Xenopus* have V_{max} values somewhat lower than those found for other anuran muscles (Table III and Fig. 6).

During growth in *Rana catesbeiana*, V_{max} decreases slightly, and the limited interspecific data are consistent with the same allometric scaling (Fig. 6). The allometric exponent is -0.094 ± 0.018 (SE), which is very similar the exponent of -0.084 found for the V_{max} of fast fibers in lizards (Marsh, 1988), and not significantly different from the value of -0.13 found for intact fast muscles and skinned fibers in mammals (Close, 1972; Seow and Ford, 1991). Why the scaling of V_{max} should be so similar among these diverse groups of animals is not known. Marsh

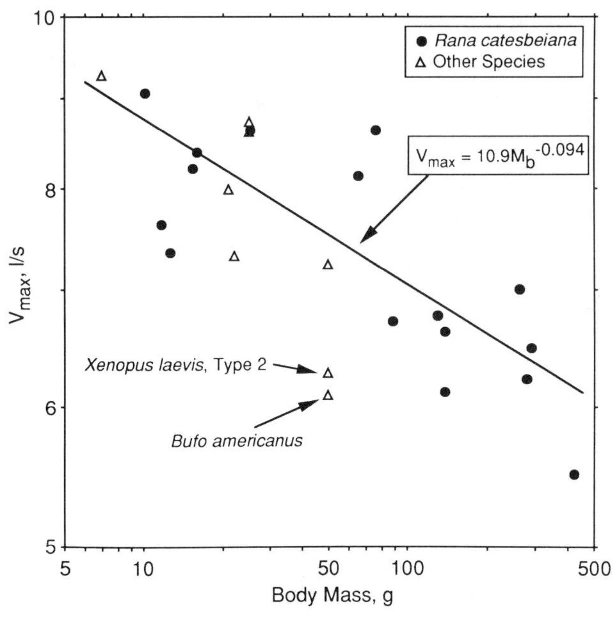

FIG. 6. Maximal isotonic velocity of shortening at 20°C as a function of body mass. Solid circles are for *Rana catesbeiana* (R. L. Marsh, unpublished); open triangles are for other species (data from Table III). Both scales are logarithmic.

(1988) pointed out that the elastic similarity hypothesis, which has been suggested as an explanation of this scaling in mammals, does not hold for running lizards. The dimensions of anurans also do not scale according to elastic similarity rules. Whatever the evolutionary explanation for the scaling of V_{max}, it has a major impact on the allometry of jumping performance (Section IV, C).

Temperature, of course, has a major influence on V_{max} (Bennett, 1984) with an overall Q_{10} of approximately 2.0 for anuran muscles (Fig. 7). The Q_{10} for V_{max} declines slightly with increasing temperature in muscles studied over a sufficiently large temperature range (Renaud and Stevens, 1984; John-Alder *et al.*, 1989) Acclimation has little or no effect on V_{max} in the species studied to date (Renaud and Stevens, 1984; Rome, 1983).

The maximum power output ($\dot{W}_{max}$) during isotonic contractions is a function of V_{max}, P_0, and the shape of the force–velocity curve, which is most conveniently expressed as the power ratio, $R_{\dot{W}}$ (Marsh and Bennett, 1986a). Few studies have reported all three of these variables, but enough data exist for some tentative conclusions to be drawn. The power ratio varies in vertebrate muscle with the fiber type studied

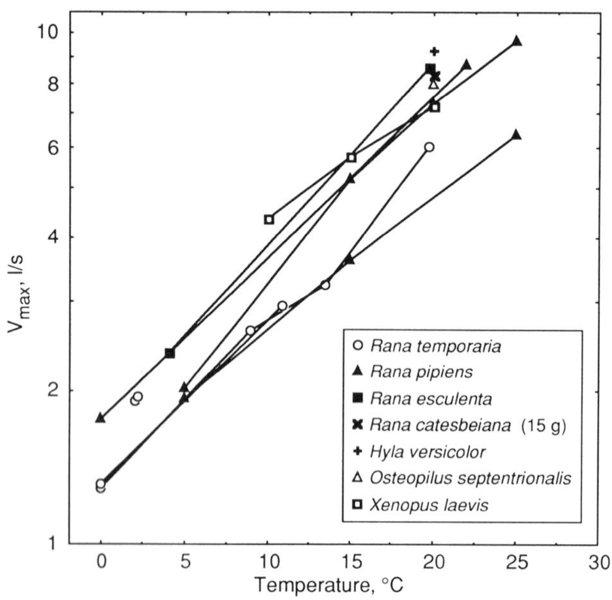

FIG. 7. Maximal isotonic velocity of shortening as a function of muscle temperature (data from Table III). The vertical scale is logarithmic.

(Woledge, 1968; Close, 1972). Among fast anuran muscles $R_{\dot{W}}$ is independent of temperature and body size (Table III) and varies from 0.08 to 0.14 with a mean of 0.113 ± 0.004 (SE). Power output of fast frog muscles studied at 20°C averages 270 ± 27 W kg^{-1} with a high value of 350 W kg^{-1} (Fig. 8). The thermal dependence of $\dot{W}_{max}$ changes with temperature, reflecting the combined effects of temperature on V_{max} and P_0 (Fig. 8). Power output of the sartorius muscle of *Rana pipiens* has also been measured at 20°C using cyclical contractions with sinusoidal length changes (Stevens, 1988). The mean power output for a complete cycle at 2 Hz was 35 W kg^{-1}, or 70 W kg^{-1} considering only the shortening phase. This latter value is only 30% of the maximum power developed during isotonic contractions.

Direct measurements of work output have been reported in numerous studies of amphibian muscle, particularly in conjunction with measurements of heat production. The goal of most of these studies was not to define the optimum conditions for work or power output, and many of them were done on muscles at 0°C, but taken together they provide a rough idea of the range of work output expected from anuran muscle. The work of Gibbs and Chapman (1974) is representative of

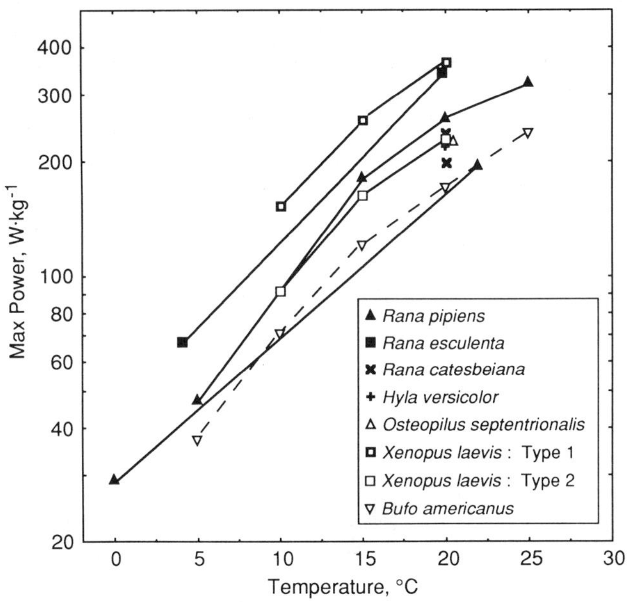

FIG. 8. Maximal power output during isotonic contractions as a function of muscle temperature (data from Table III). The vertical scale is logarithmic.

studies employing isotonic loading. They examined isotonic work output at 10° and 20°C, temperatures at which measurements of jumping performance are also available. The maximum work output during tetanic contractions was 29 and 33 J kg^{-1} at 10°C and 34 and 40 J kg^{-1} at 20°C for sartorii from the frog *Litorea aurea* and the toad *Bufo marinus*, respectively. Early experiments of A. V. Hill and colleagues with inertial levers probably come the closest to the loading conditions that prevail during jumping. Using this system a maximum of 28 J kg^{-1} was recorded for the sartorius of *Rana temporaria* at 21°C (Hill and Hartree, 1920). This study did not explore a full range of inertial loads and Hill (1922) suggested that frog muscle should be capable of 60 J kg^{-1} when contracting against very heavy loads. Maximal work per cycle is 17 J kg^{-1} during 2-Hz sinusoidal length cycles at 20°C with a strain of $0.12L_0$ (Stevens, 1988).

B. Muscles Used in Jumping and Their Dimensions

Before examining the amount of muscle available to power the jump, the muscles that could be involved in this movement need to be identified. A comprehensive treatment of the anatomy of anuran hindlimb muscles is found in Duellman and Trueb (1986). These authors follow the nomenclature of Dunlap (1960) and include a description of the major actions of the muscles as deduced from their anatomy. Matsui (1978) also provides a useful diagrammatic summary of the lines of action of the hindlimb muscles. Many of the muscles used during jumping cross two joints and their precise actions during jumping will depend on their moment arms and the rate of rotation of each joint. Calow and Alexander (1973) attempted to take these factors into account and listed the following six major extensor muscles as providing power for the jump:

1. m. plantaris longus (gastrocnemius): extends the ankle; small flexor moment at the knee;
2. m. curalis (part of triceps femoris): extends the knee; small flexor moment at the hip;
3. m. gluteus magnus (part of triceps femoris): extends the knee; larger flexor moment at the hip than the curalis;
4. m. semimembranosus: extends the hip (femoral adduction according to Duellman and Trueb, 1986); very small flexor moment at the knee;
5. m. gracilis major: extends the hip; smaller flexor moment at the knee;

6. m. semitendinosus: extends the hip; flexes the knee (net action unclear).

The function of the semitendinosus during jumping remains unclear. Calow and Alexander (1973) estimate that it it shortens during the jump. However, Mai and Lieber (1990) use additional anatomical measurements and the joint angles measured by Calow and Alexander (1973) to conclude that the ventral head of the semitendinosus lengthens throughout most of the jump. Electromyographic (EMG) recordings indicate that the semitendinosus is active in some jumps and not others (J. M. Olson and R. L. Marsh, unpublished).

This list of muscles that could power the jump probably should be supplemented with one major and a number of minor hindlimb muscles.

1. m. adductor magnus: adducts the femur;
2. m. adductor longus: adducts the femur;
3. m. tibialis posticus: similar action to plantaris;
4. m. peroneus: extends the knee;
5. m. tensor fasciae latae (part of triceps femoris): similar action to curalis but larger flexor moment at hip;
6. m. gracilis minor: similar action to gracilis major;
7. m. obturator externus: extends the hip;
8. m. quadratus femoris: adducts the femur;
9. m. gemellis: adducts the femur.

The adductor magnus is one of the larger muscles in the hindlimb (Matsui, 1978) and it would be surprising if it did not help power jumping. As the hindlimb is moved from the prejump to the takeoff position this muscle clearly shortens, in part due to the rotation of the femur (J. M. Olson and R. L. Marsh, unpublished). Electromyograms indicate that this muscle is active concurrently with the other major leg muscles during jumping (Fig. 9). The three deep thigh muscles listed here, the obturator externus, the quadratus femoris, and the gemellis, have potential actions that could contribute to jumping, but their small size precludes a major role. Several muscles in the lower back that extend the iliosacral joint are also active during the jump and could contribute to powering the jump (Emerson and De Jongh, 1980). However, the contribution of these back muscles to total power output is probably small because taken together they amount to only about 0.5% of body mass.

Data on the mass of the muscles involved in jumping are crucial to integrating intrinsic contractile properties with jumping performance.

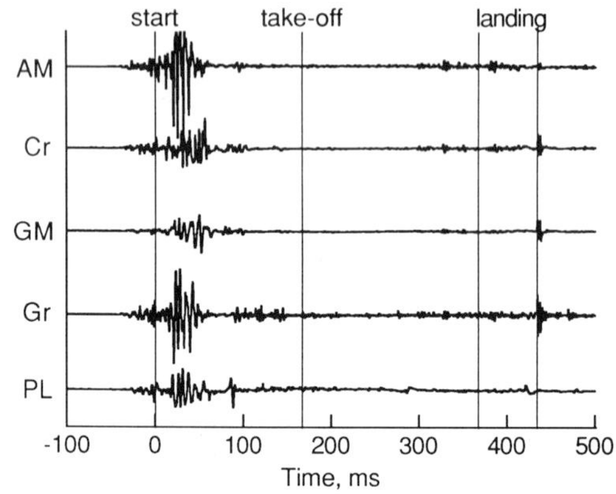

FIG. 9. Electromyographic recordings from five muscles during jumping in *Rana catesbeiana*. The times of the "start" of the jump, "takeoff," and "landing" were determined by simultaneous video recordings at 60 Hz. AM, adductor magnus; Cr, curalis; GM, gluteus magnus; Gr, gracilis major; PL, plantaris longus.

Unfortunately, only a few measurements of hindlimb muscle mass are available for anurans (Table V). In ranids, which are in general strong jumpers, as much as 20% of the body mass is devoted to muscles that could potentially power the jump. Muscle mass is lower in two hylids for which data are available. Of the species measured, *Bufo* has the smallest percentage of body mass devoted to hindlimb muscles (Putnam, 1979) and also has a smaller percentage of this muscle mass devoted to extensors and femoral adductors (Matsui, 1978).

The data provided here are too few to allow any conclusions regarding the interspecific allometry of the mass of the muscles that power the jump, although the data on *Hyla* (Table V) are suggestive of relatively smaller muscles in smaller frogs. Emerson (1978) provides data on the intraspecific allometry of the combined mass of the six extensor muscles specified by Calow and Alexander (1973). She found the following allometric exponents based on dry muscle masses of museum specimens: *Rana catesbeiana*, 1.12; *Rana erythracea*, 1.03; *Bufo asper*, 1.11; and *Hyla regilla*, 0.84. A smaller number of measurements of fresh muscle mass in *Rana catesbeiana* yield an exponent of 1.08 (J. M. Olson and R. L. Marsh, unpublished), confirming a slight positive allometry of muscle mass during growth. Intriguingly, a slight positive allometry also exists during growth in lizards (Garland, 1984, 1985) and interspecifically in mammals (Alexander *et al.*, 1981).

Even fewer data are available from which to judge the allometry of the length and cross-sectional area of the hindlimb muscles. Emerson

TABLE V

MASS OF THE MAJOR HINDLIMB MUSCLES IN ANURANS AS A PERCENTAGE OF TOTAL BODY MASS

Species	M_b (g)	A: Six major extensors[a] ($\%M_b$)	B: Other extensors and femoral adductors[b] ($\%M_b$)	A + B: $\%M_b$	Total hindlimb muscle mass $\%M_b$	Reference
Rana temporaria	23	10		14[c]	17	Calow and Alexander (1973)
Rana pipiens	~25	14		19[c]	23[c]	Hirano and Rome (1984)
Rana pipiens	64			20[c]	24	Putnam (1979)
Rana catesbeiana	325	14.7	4.3	19	24	J. M. Olson and R. L. Marsh (unpublished)
Pseudacris crucifer	1				18	Taigen *et al.* (1985)
Hyla versicolor	8				14	Marsh and Taigen (1987)
Bufo boreas	59			9[c]	12	Putnam (1979)

[a] Extensors listed by Calow and Alexander (1973).
[b] Other muscles that could power the jump; see text.
[c] Estimates based on the relative muscle proportions in *Rana* and *Bufo* given by Matsui (1978).

(1985) has suggested that positive allometry of muscle cross-sectional area could increase the "specific energy" available in larger animals. Maximum work and power are, however, related to muscle mass and not cross-sectional area. With a given mass of muscle, increasing the cross-sectional area, for example, by increasing the angle of pinnation, will decrease the distance shortened and result in no change in maximum work output. As a first approximation one might expect that muscle length would scale proportional to the length of the hindlimbs (L_{hl}). This expectation is confirmed during growth in *Rana catesbeiana*, in which several length measures scale as $M_{\mathrm{b}}^{0.3}$ (Table IV). Coupled with the slight positive allometry of muscle mass, these data suggest that cross-sectional area scales as $M_{\mathrm{b}}^{0.8}$ intraspecifically. Somewhat puzzling in this context are the data of Sperry (1981). He calculated cross-sectional area from counts of fiber number and samples of fiber cross-sectional area. He found that predicted cross-sectional area scales as approximately $M_{\mathrm{b}}^{0.9}$ in *Rana pipiens* and *Bufo americanus*. Considering that length scales as $M_{\mathrm{b}}^{0.3}$, Sperry's data would suggest a considerably larger allometry of muscle mass than has actually been measured.

IV. Jumping Performance

The most common measure of performance available for anurans is jumping distance (Wermel, 1934; Rand, 1952; Rand and Rand, 1966; Zug, 1972, 1978, 1985; Dobrowolska, 1973; Zug and Altig, 1978; Emerson, 1978; Renaud and Stevens, 1983b; Londos and Brooks, 1988, 1990; John-Alder *et al.*, 1988; Whitehead *et al.*, 1989; John-Alder and Morin, 1990; Knowles and Weigl, 1990). Smaller amounts of data are available on takeoff velocities, ground reaction forces, and metabolic support for jumping (Hirsch, 1931; Calow and Alexander, 1973; Hirano and Rome, 1984; Anderson *et al.*, 1991; Wine and Gatten, 1992). Fortunately, because of the simple mechanics of jumping, jumping distance can be related to mechanical work and power (Section II, C).

Summarizing the available data on jumping distance is complicated because a variety of experimental protocols and several measures of jumping performance have been used. Because the primary metabolic support for jumping in many frogs is anaerobic, studies that average large numbers of sequential jumps may underestimate jumping capacity as a result of fatigue (Zug, 1985). Other studies have reported mean jumping distance ($L_{\mathrm{j,ave}}$) in a smaller number of jumps, for example, 10. However, most jumps undertaken in experimental settings are submaximal and this measure probably also underestimates maximal performance. An alternative is to report the maximum jump recorded for any individual of the species studied ($L_{j,record}$). This value, although of interest, may include experimental errors that bias it toward longer jumps. An alternative estimate of maximal performance employed by several investigators is to record the maximum jumping distance for each individual and average these values to obtain a mean maximal jumping distance ($L_{j,avemax}$). This value may underestimate maximal capacity somewhat, but probably provides a reliable value to use in examining the effects of body size, temperature, and other variables.

A. Interspecific Variation in Jumping Performance Including Allometry

The most complete study of jumping performance in anurans is that of Zug (1978). Because variation in methods can increase the already large inherent variability in behavioral measures, this study forms an important benchmark. Zug recorded data on 81 species soon after capture with similar methods. The body form among anurans varies considerably and some consideration must be given to this variation in discussing the interspecific differences found by Zug (1978). Other studies have documented the influence of limb length and other variables on jumping distance (Wermel, 1934; Zug, 1972; Emerson, 1978), but these variables are not completely analyzed in Zug's study. However, some variation in body form is separable at the familial level, which is probably the finest division that is useful in this post hoc analysis. Past studies have shown, for example, that toads in the family Bufonidae have poor jumping performance compared with species usually called frogs, for example, ranids. In the following analysis, I have grouped the pelobatid (spadefoot) toads from North America with the bufonids. These two families are similar in overall morphology and locomotor behavior and are referred to as "Toads." Dividing the data at the familial level creates some problems because of the diversity of body form and behavior that exists within many anuran families. Intrafamilial variation in body form is particularly striking among the four species of microhylids studied by Zug (1978). Also, the dendrobatid frogs studied by Zug include two species of "poison-arrow frogs" that jump poorly compared with other members of this family. Rather than proposing criteria for excluding individual species, all microhylids and dendrobatids were removed from the analysis of covariance discussed in the following. (Inclusion of these groups increases the variability in the data without affecting the conclusions.)

A useful first approach to broad interspecific comparisons is allometric analysis. This analysis allows the pervasive effects of body size to be described and, importantly, allows the comparison of empirically and theoretically derived allometric exponents. For this analysis Zug's sample was divided into three groups for analysis of covariance (ANCOVA) after first excluding species for which three or fewer individuals were tested. First, members of the families Bufonidae (true toads) and Pelobatidae (spadefoot toads) were grouped as "Toads." With one exception all toads were jumped at an air temperature of approximately 23°C. The second group, "Frogs-23C," consisted of frogs also jumped at approximately 23°C. This group includes members of the Ascaphidae, Hylidae, Leptodactylidae, and Ranidae. The third group, "Frogs-29C," were frogs from Papua New Guinea that were jumped at air temperatures from 25° to 34°C, with most trials occurring at 28°–31°C. This group consisted of frogs from the families Myobatrachidae, Hylidae, and Ranidae.

All measures of jumping performance show a significant and consistent effect of body size interspecifically (Fig. 10 and Table VI). For $L_{j,avemax}$ or $L_{j,record}$ the common slope was 0.20 ± 0.02 (SE) and for $L_{j,ave}$ this value was 0.19 ± 0.03. Based on comparing the intercept values, $L_{j,ave}$ is about 15% lower than $L_{j,avemax}$, whereas $L_{j,record}$ is about 15% higher than this value.

The frogs tested at high temperatures by Zug show clearly enhanced performance. The intercepts for the Frog-29C group are approximately 1.5 times the values for the Frog-23C group (Table VI). If this difference were solely due to temperature and the body temperatures (T_b) of the frogs were equal to the air temperature at which they were jumped, it would reflect an R_{10} of 2.5. This level of thermal dependence is greater than has been found intraspecifically over this temperature range (Section IV, E). However, the interspecific thermal dependence may be higher, particularly when comparing temperate and tropical species. Temperature shifts in performance curves are a common feature of the evolutionary adaptations of ectotherms (Bennett, 1990). Also, because of the high water permeability of amphibian skin, the T_b of frogs at a given air temperature will vary with the relative humidity. Thus, if the relative humidity was higher during the tests in New Guinea, the difference in T_b of the two groups tested by Zug may have been greater than indicated by the air temperatures. Another factor in determining the apparently high R_{10} may be differences between the groups in relative size of the hindlimbs. The best jumpers in Zug's study were New Guinean hylid tree frogs in the subfamily Pelodryadinae. This group includes the individual with the longest jump recorded, 2.27 m by an "Australian rocket frog" (*Litoria nasuta*), which is known for using a series of powerful jumps to escape predation (Tyler *et al.*, 1984). Frogs in the genus *Litoria* appear to have somewhat longer hindlimbs relative to body size than do the North American hylids or ranids tested by Zug (Tyler *et al.*, 1984).

Jumping distance varies considerably among the species studied by Zug. Some of this variation certainly represents unavoidable variability associated with behavioral measurements, but the differences may also provide useful hints for examining the correlates of performance in anurans. The jumping performance of the bufonid and pelobatid toads averages only 36% of that found for frogs (Table VI and Fig. 10), and morphological and physiological factors can explain this difference (Section IV, D). The toadlike mycrohylids also jump relatively poorly. The physiological basis for the poor jumping performance of the two poison-arrow frogs is less clear, but obviously, the diminished jumping performance in these highly toxic frogs could relate to a relaxation of selection on predator avoidance behaviors. Zug recorded exceptionally long jumps by *Litoria nasuta* and the cricket frog, *Acris gryllus* (see also Rand, 1952; John-Alder *et al.*, 1988; Knowles and Weigl, 1990)

TABLE VI

INTERSPECIFIC ALLOMETRY OF JUMPING DISTANCE BASED ON THE DATA OF ZUG (1978)[a]

Variable	Group	r	a	b	95% confidence interval
$L_{j,ave}$		0.875		0.191	0.131–0.251
	Frog-23C		0.337		
	Frog-29C		0.562		
	Toad		0.132		
$L_{j,avemax}$		0.912		0.200	0.154–0.246
	Frog-23C		0.418		
	Frog-29C		0.608		
	Toad		0.153		
$L_{j,record}$		0.88		0.200	0.150–0.250
	Frog-23C		0.499		
	Frog-29C		0.721		
	Toad		0.166		

[a] After exclusion of the microhylids and dendrobatids, the data were divided into three groups (see text) and subjected to ANCOVA after log transformation. The slopes of the individual groups were not significantly different for any variable, and the common slope determined by ANCOVA is presented. Coefficients are presented for the allometric equation $L_j = aM_b{}^b$(L_j in m and M_b in g).

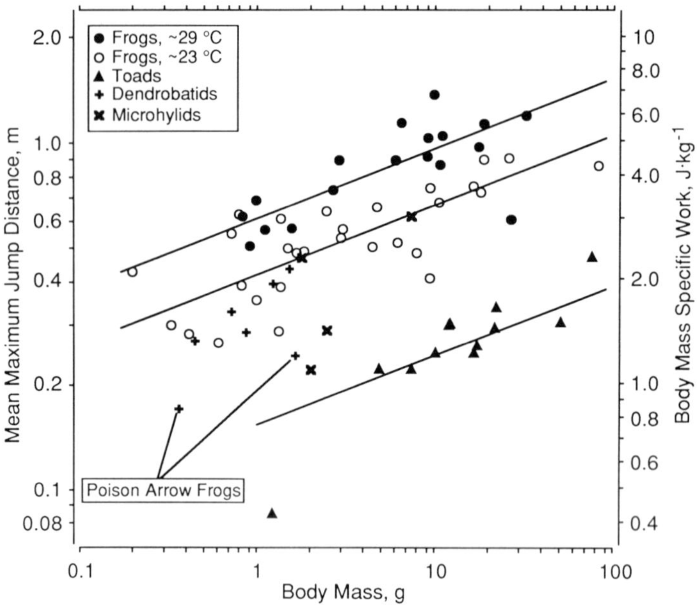

FIG. 10. Mean maximal jumping distance and body-mass-specific work in anurans (data from Zug, 1978). See also Table VI. Both scales are logarithmic.

B. INTRASPECIFIC ALLOMETRY

Several studies have documented the effects of body size on jumping performance intraspecifically (Emerson, 1978; Zug, 1978). Unfortunately, some of these studies implicitly assumed that jumping distance is directly proportional to snout–vent length. This assumption led Zug (1978) to use relative jumping distance (jumping distance/snout–vent length) in many comparisons, under the assumption that with this measurement, "the size factor can be removed from the actual jumping distance" (Zug, 1978). A better initial assumption might be that intraspecific and interspecific allometry will be the same. As shown earlier, jumping distance scales interspecifically approximately as $M_b^{0.2}$. Zug (1978) provides allometric equations relating jumping distance to snout–vent length in 27 species. Because of the relatively small range of body sizes in these intraspecific comparisons and the variability in jumping performance, the allometric exponents are quite variable. However, the mean exponents relating $L_{j,ave}$ and $L_{j,avemax}$ to snout–vent length are 0.62 and 0.57, respectively. These slopes predict a body mass exponent of approximately 0.2. A similar estimate of intraspecific allometry of jumping distance is obtained by averaging the slopes for 7 species for which Zug provides direct data on the allometry with body mass. Three species studied by Emerson (1978) show comparable allometric exponents (0.18–0.31). Overall, these data are consistent with the hypothesis that jumping distance scales similarly both intra- and interspecifically. Some of the variation in allometric slopes among various species may be hypothesized to indicate differences in the underlying allometry of power output (Section IV, C), but the data necessary to evaluate this hypothesis are not available for most species.

Apparently, the jumping performance of an anuran during much of postmetamorphic growth is predictable from body size effects that also pertain interspecifically. However, in the period immediately following metamorphosis, jumping distances are shorter than expected (Emerson, 1978; Zug, 1978). The reasons for the change in allometry are unclear, but further study of this phenomenon might provide useful clues regarding the physiological correlates of performance.

C. Predicted Allometric Scaling

Hill's (1950) prediction that all animals should run the same speed and jump the same distance regardless of their size has often been quoted, including in the literature on jumping in frogs (Zug, 1978; Emerson, 1978). Hill's speculative conclusions have at times been uncritically accepted as a precise prediction without examining the validity of his underlying assumptions. Also common in the literature is the mistaken impression that the allometry predicted by Hill is a consequence of "geometric" scaling of animals' dimensions per se. Instead, the predictions stem from assumptions about the scaling of dimensions coupled with assumptions about the scaling of the intrinsic contractile properties of muscle. Specifically, Hill's conclusions derive from the following assumptions: (1) the dimensions of different-sized animals follow the principles of geometric similarity; (2) the force per cross-sectional area of skeletal muscle is independent of size; and (3) the intrinsic shortening velocity of skeletal muscle (V_{max} in L_0 s^{-1}) scales as $M_b^{-1/3}$. The assumed scaling of V_{max} combined with the geometric scaling of muscle length ($L_0 \propto M_b^{1/3}$) leads to the the prediction that maximum shortening velocity in absolute units, that is, mm s^{-1}, will be independent of size. Hill made these assumptions in the context of the further assumption that animals will move in ways that keep inertial forces constant (McMahon, 1984; Marsh, 1988). Certainly not incidentally, this set of assumptions leads to a simple and internally consistent set of predictions about the way animals should move. In the case of jumping, Hili's assumptions predict size-independent jumping distance on the basis of either Eq. (6) or Eq. (18), relating jumping distance to takeoff velocity or power output by the muscles. The prediction follows from either equation if takeoff angle is constant, muscle shortening velocity is proportional to takeoff velocity, and the muscles shorten at constant fractions of V_{max} and P_0 during takeoff (power = force × velocity).

Unfortunately, several of Hill's assumptions have turned out not to be true. As reviewed earlier, the following proportionalities are the best current estimates of the allometry of the relevant characteristics of anurans.

Muscle mass (M_m) or volume $\propto M_b^{1.1}$
Muscle length (L_0) $\propto M_b^{0.3}$
Hindlimb length (L_{hl}) $\propto M_b^{0.3}$
Muscle cross-sectional area ($A_m \propto M_m/L_0$) $\propto M_b^{0.8}$
Intrinsic muscle strength (P_0, N m^{-2}) $\propto M_b^0$
Intrinsic muscle shortening velocity (V_{max}, L_0 s^{-1}) $\propto M_b^{-0.1}$
Absolute muscle shortening velocity (mm s^{-1}) $\propto V_{max}\, L_0 \propto M_b^{0.2}$
Muscle power ratio ($R_{\dot{W}} = \dot{W}_{max}/V_{max}\, P_0$) $\propto M_b^0$

The jumping performance of an animal will depend on the mechanical power transferred to the center of mass during takeoff. Therefore, a reasonable starting point for predicting the allometry of jumping is Eq. (18), which relates jumping distance to the power output and the linear dimensions of the hindlimbs. The preceding proportionalities predict that maximum power output per unit body mass in anurans will be independent of body mass.

$$\dot{W}_{max}\, M_b^{-1} \propto P_0\, A_m\, V_{max}\, L_m\, M_b^{-1} \propto M_b^0.$$

The slight negative allometry of the length-specific intrinsic shortening velocity is offset by the slight positive allometry of muscle mass. Therefore, although muscle-mass-specific power output is expected to decline because of decreasing V_{max}, body-mass-specific power output is predicted to remain constant. If one assumes that the power output in the jump is proportional to the maximum isotonic power output of the muscles, then Eq. (18) predicts that jumping distance will be proportional to the $\frac{2}{3}$ power of hindlimb length, or $M_b^{0.2}$.

Given the actual scaling of anuran morphology and muscle contractile properties and the further assumption that takeoff velocity will be proportional to V_{max}, Eq. (6) does not predict the same allometry of jumping distance as Eq. (18). However, the takeoff velocity is the result of the acceleration of the body mass, and thus is dependent on the power output of the muscles. Using Eq. (18) amounts to a prediction of takeoff velocity on the basis of maximal muscle power output. This assumption that the actual power output will be proportional to the maximum isotonic power might be criticized because the predicted takeoff velocities based on Eq. (18) are not proportional to V_{max}. The muscles in different-sized animals thus might be expected to operate at different points on their force–velocity curves and produce different fractions of peak isotonic power (Fig. 5). Two further considerations may be used to temper this criticism. First, the power curve as a function of either velocity or force is quite broad. Using a typical anuran force–velocity curve, power output is within 80% of the maximum value as V/V_{max} is changed from 0.14 to 0.49 (Fig. 5). Second, the diverse muscles used to power the jump almost certainly do not operate isotonically during jumping. However, the lack of detailed knowledge of the actual dynamics of the major muscles during jumping (Section V, C) precludes a precise evaluation of their ability to produce power. One possible assumption is that these muscles operate over a region of the force–velocity curve that is determined by the acceleration of a mass. In fact, Calow and Alexander (1973) have previously argued that the dynamics of the anuran muscles during jumping should be considered in this context.

The allometric consequences of Calow and Alexander's model were determined by running a similar model (Marsh, 1990) and varying the accelerated mass. The dimensions and maximal shortening velocity of the muscle in the model were simultaneously altered as prescribed by allometric rules. If the model is run with the allometry assumed by Hill (1950), all of Hill's initial predictions remain unchanged (Fig. 11). The final velocity is independent of accelerated mass; mass-specific work remains constant; and mass-specific power output declines with a mass exponent of −0.33. However, if the empirically derived allometric rules are followed, the predicted performance matches more closely the actual performance during jumping (Fig. 11). Final shortening velocity is predicted to be proportional to mass to the 0.13, despite the fact that maximal muscle shortening velocity scales as $mass^{0.2}$. The empirically derived allometry also predicts that the mass-specific work output will increase substantially with increasing mass. Therefore, the predicted mass-specific power output varies much less than predicted by Hill. The difference between the performance predicted by Hill's assumptions and that predicted from empirically derived allometric rules stems mostly from the difference in the scaling of V_{max}. If the model is run with geometric scaling of dimensions, but

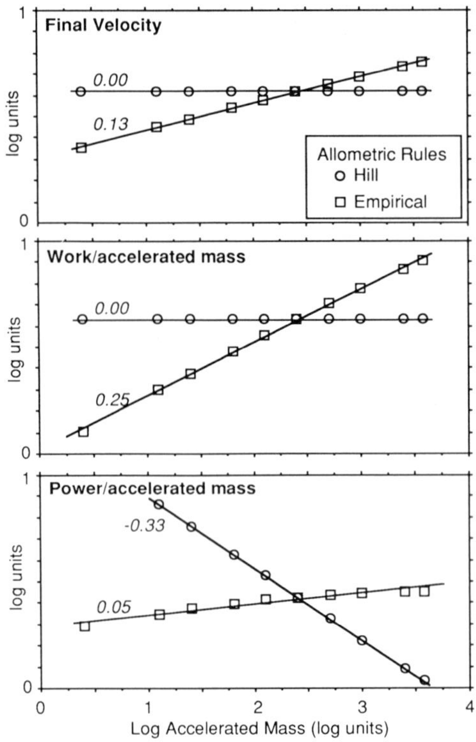

FIG. 11. Allometry of final shortening velocity, work per unit mass, and power per unit mass based on a model of a muscle starting at P_0 and shortening against an inertial mass (Calow and Alexander, 1973; Marsh, 1990; see text). Both scales are logarithmic.

V_{max} scaling as $M_b^{-0.1}$, then final velocity scales as mass$^{0.11}$. This model is admittedly very simple and does not incorporate the length-dependent changes in force and variable activation that almost certainly occur in actual jumping. Even so, the results indicate that incorporating the characteristics of the force–velocity curve does not necessarily alter our predictions of the allometry of jumping performance.

The conclusion from the foregoing analysis is that jumping distance is not expected to be mass independent as suggested by Hill (1950) and others. The positive allometry of jumping distance is largely due to a large difference between the actual scaling of intrinsic shortening velocity ($M_b^{-0.1}$) and that assumed by Hill ($M_b^{-0.33}$). However, this analysis clearly does not address why natural selection has produced the observed allometry of intrinsic velocity. As reviewed in Section III,A, the scaling of V_{max} in anurans is quite similar to that found in lizards (Marsh, 1988) and in mammals (Close, 1972; Seow and Ford, 1991). The selective factors that join the diverse locomotor patterns found in these groups remain to be determined.

D. Work and Power Output

The body-mass-specific work output during jumping can be derived from jumping distance assuming that the frogs jumped at close to the optimum takeoff angle (Fig. 10). Based on $L_{j,avemax}$, the body-mass-specific work output of frogs jumping at approximately 23°C is predicted to increase from 2.05 J kg^{-1} for a 1-g frog to 5.14 J kg^{-1} for a 100-g frog. Converting these values to mass-specific work by the muscles requires knowledge of the fraction of body mass devoted to the muscles that power the jump, and this knowledge is slim for anurans (Table V). In ranid and hylid frogs these muscles probably occupy approximately 15–20% of M_b. Using these values, the predicted muscle-mass-specific work output (26–34 J kg^{-1} for a 100-g frog) is within the capacity for work output estimated from fast anuran muscles measured *in vitro* at 20°C (28–34 J kg^{-1}; Hill and Hartree, 1920; Gibbs and Chapman, 1974).

The preceding allometric analysis predicted that body-mass-specific power output in anurans should be independent of body mass. The power output required to produce the jumps recorded by Zug (1978) can be calculated from jumping distance, M_b, and L_{cm}, which can be estimated as 1.15 L_{hl} (Section II,C). Information is available on L_{hl} as a fraction of L_{sv} for some individual species used by Zug (Rand, 1952; Zug and Altig, 1978). Less precise, but reasonable, estimates of L_{hl} can be made on the basis of measurements of similar species in the same family.

The predicted power outputs for frogs are impressively high (Fig. 12 and Table VII). Using the data for the $L_{j,avemax}$, the predicted body-mass-specific power output for frogs tested at an air temperature of 23°C is about 50 W kg^{-1} body mass, and several individual species produce power in the 70–80 W kg^{-1} range (Fig. 12A). These high power outputs have been confirmed by direct measurements based on filming (Table II). Considering that at most 20% of the body mass consists of the muscles that actually power the jump (see Table V), these muscles appear to be able to produce at least 250–400 W kg^{-1} during jumping at a temperature of approximately 23°C. Even higher estimates are obtained if the longest jumps recorded by Zug (1978) are used in the calculations (Fig. 12B and Table VII). Individual frog Type 1 fibers, or

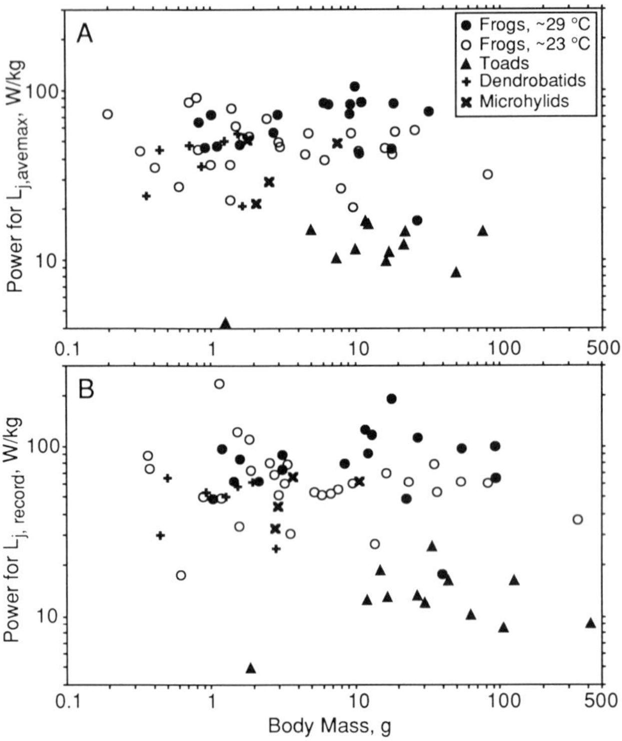

FIG. 12. Average mechanical power output during jumping in anurans (calculated from jumping distances in Zug, 1978). Both scales are logarithmic. See also Table VII.

muscles consisting of mostly this fiber type, produce an average maximum power of about 270 W kg^{-1} during *in vitro* isotonic contractions at 20°C (Table III and Fig. 8). Allowing for a Q_{10} of 2, these muscles should be able to produce 330 W kg^{-1} at 23°C. Even if one assumes that the highest values recorded *in vitro* are closer to the actual capabilities of these muscles, frog muscle ought to be able to produce a maximum power of about 430 W kg^{-1} at 23°C based on isotonic measurements.

The jumping performance of toads also indicates that their muscles perform well in relation to *in vivo* measurements (Table VII and Fig. 12). Based on $L_{j,avemax}$, toads produce an average of 12 W kg^{-1} body mass, with values for individual species as high as 16 W kg^{-1}. Again, even higher values are estimated if the data on record jumps are used. The muscles that produce power for jumping probably amount to 9%, or less, of body mass in toads (Table V). Thus, these muscles appear to be able to produce at least 130–180 W kg^{-1} of power during jumping at 23°C. Measurements on the sartorius muscle of *Bufo americanus* indicate that this muscle can produce a maximum isotonic power of approximately 210 W kg^{-1} at 23°C (Fig. 8).

Approaching jumping performance from the perspective of muscle power production also suggests that interspecific differences in jumping performance between frogs and toads can be partitioned into three main effects. The expected maximum power output of the hindlimb muscles in toads is only 25% of that found in ranid frogs because of two factors: (1) a 37% reduction in muscle-mass-specific power output due to intrinsic differences in the muscle fibers and (2) a 60% reduction in muscle mass. Toads also have hindlimbs that are about 20% shorter than those of ranid frogs of comparable size. Equation (18) thus predicts that toads should only jump 34% as far as frogs. This estimate matches closely the value estimated by analysis of covariance, which indicates that toads jump on average 37% of the distance jumped by the same size frog (Fig. 10 and Table VI).

The data on jumping distance would, therefore, suggest that anuran muscle can produce an average power during jumping that is at least equal to its maximum isotonic power. This conclusion is surprising

TABLE VII

CALCULATED AVERAGE POWER REQUIRED TO PRODUCE THE JUMPING DISTANCES MEASURED BY ZUG (1978)[a]

	Group	Predicted power in jumping (W kg^{-1})
$L_{j,ave}$		
	Frog-23C	36.1 ± 2.7
	Frog-29C	56.9 ± 4.2
	Toad	8.7 ± 0.8
$L_{j,avemax}$		
	Frog-23C	49.3 ± 3.3
	Frog-29C	66.3 ± 5.1
	Toad	12.1 ± 1.0
$L_{j,record}$		
	Frog-23C	67.3 ± 7.4
	Frog-29C	87.6 ± 8.9
	Toad	13.4 ± 1.5

[a] The calculated power is the average value for the entire takeoff phase. Peak powers would be approximately double these values (see text). Valuse are presented as means ± SE and expressed per kilogram of body mass.

given that the jump is powered by a diverse array of muscles with different actions, and undoubtedly different patterns of shortening as well (Section V,C). Noteworthy is the fact that the power output calculated from jumping distance represents the power averaged over the entire contact time and peak power during the jump is expected to be about twice as great (Sections V,A and V,B) (Marsh and John-Alder, 1994).

Power output during jumping could be enhanced if energy was stored prior to the jump, using a mechanism analogous to those found in some insects (Bennet-Clark, 1977). In this mode of jumping, energy is stored in elastic structures prior to the jump, and then released by elastic recoil during takeoff. The storage and release of energy allows power to be produced during takeoff at rates that would be impossible from direct contraction of the muscles. To increase the *average* power during takeoff, a "catch" is required to prevent movement during the initial phase of contraction. During jumping in frogs, muscle activity does begin before movement is detected (see Figs. 9 and 16), but the muscles are also deactivated before takeoff. Thus, the total period of muscle activity is not longer than the contact time. However, storage of elastic energy could still help redistribute power output during jumping and thus increase the *peak* power. Peak power output occurs late in the takeoff phase (see Fig. 15). Energy stored in series elastic structures, for example, tendons, early in the jump could be released as the frog accelerates, and only a portion of the peak power would have to be supplied directly by the muscles.

E. Thermal Effects

Temperature has pervasive effects on biological processes and ecologists and physiologists have devoted considerable effort to describing the effects of temperature on locomotor performance (Bennett, 1990). The bulk of the data on thermal effects on jumping in anurans conforms to one basic pattern (Table VIII and Fig. 13). Jumping distance is thermally dependent, but with a declining R_{10} as temperature increases. Individual species do differ markedly in absolute jumping performance and in the ability to jump well at low T_b (John-Alder *et al.*, 1988). Acclimation to cold in the laboratory does not improve jumping performance of anurans at cold temperatures (Renaud and Stevens, 1983b; Londos and Brooks, 1988; Knowles and Weigl, 1990). On the other hand, the same studies have noted that cold-acclimated frogs and toads do not perform as well as warm-acclimated individuals when tested at warm temperatures. The biological significance of this differ-

TABLE VIII

THERMAL DEPENDENCE (R_{10}) OF JUMPING DISTANCE BASED ON 6 STUDIES OF 17 SPECIES OF ANURANS[a]

n	Midpoint T_b (°C)	R_{10} of jump distance		Q_{10} of power output	
		Total	Aerial	Jumping	Maximum isotonic
14	7.8 ± 0.11	2.15 ± 0.21	2.62 ± 0.37	4.24	3.4
15	12.7 ± 0.16	1.59 ± 0.09	1.73 ± 0.12	2.28	3.1
20	17.6 ± 0.14	1.34 ± 0.06	1.42 ± 0.07	1.69	1.9
19	22.6 ± 0.13	1.32 ± 0.04	1.40 ± 0.06	1.66	1.8
17	27.3 ± 0.16	1.04 ± 0.05	1.04 ± 0.06	1.06	
10	32.1 ± 0.30	0.96 ± 0.14	0.96 ± 0.15	0.94	

[a] See Fig. 17. Aerial jump distance was calculated from jumping distance and the length of the hindlimbs assuming a takeoff angle of 40° (see text). The Q_{10} of power output by the muscles during jumping was calculated from the R_{10} of aerial jumping distance based on Eq. (15). The approximate Q_{10} of maximum isotonic power is based on measurements of muscles from three species, *Rana pipiens, Xenopus laevis,* and *Bufo americanus,* in which power output of the muscles has been measured *in vitro* at closely spaced temperature intervals (see Table III).

ence is difficult to interpret. No information is available on jumping performance of naturally acclimatized individuals, and maintaining captive animals in good health during cold acclimation is problematical (Knowles and Weigl, 1990).

In contrast to this general pattern of thermal dependence, two studies report thermally independent jumping performance over a very broad range of temperatures (Huey and Stevenson, 1979; Tracy, 1979). Huey and Stevenson's report was based on a single individual and was meant to be illustrative rather than definitive. Tracy (1979) measured jumping performances that were clearly submaximal at warmer temperatures. Other studies on the same species have failed to confirm the findings of these investigators (Renaud and Stevens, 1983b; Hirano and Rome, 1984; Knowles and Weigl, 1990).

Some discussions of thermal effects on anuran jumping have left the impression that these effects are difficult to explain on the basis of the underlying properties of the muscles (Whitehead *et al.*, 1989). Understanding the basis for the thermal performance curves of anuran jumping has been hampered by an incomplete analysis of the relation between muscle kinetics and jumping distance. Jumping distance is expected to depend on the power output of the muscles, but this relation is not linear. Equation (18) predicts $L_j \propto \dot{W}_j^{0.67}$, or $\dot{W}_j \propto L_j^{1.5}$. Addi-

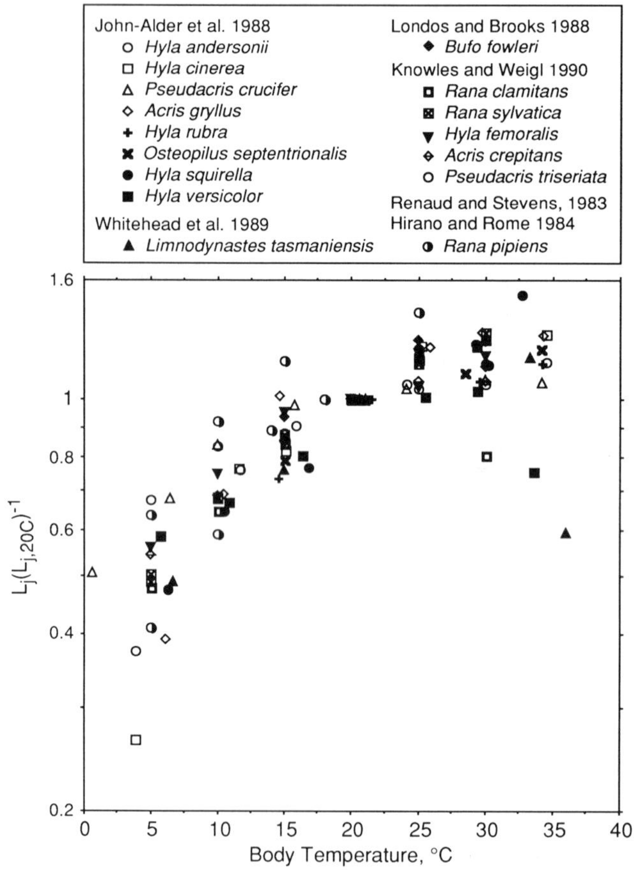

FIG. 13. Jumping distance as a function of body temperature in anurans. Distance is expressed relative to the value measured at 20°C. The vertical scale is logarithmic.

tionally, the power expended during takeoff predicts the aerial jumping distance and not the total jumping distance (Section II,C). Comparing the R_{10} of jumping distance to the Q_{10} of muscle power output therefore requires two corrections. First, the $L_{j,a}$ should be estimated based on the dimensions of the animals. Then the thermal dependence of mechanical power can be calculated as the R_{10} of $L_{j,a}$ raised to the 1.5 power.

With these corrections, the estimated thermal dependence of muscle power output during jumping is quite high (Table VIII). The trend toward higher Q_{10}'s for power output at lower temperatures occurs both *in vivo* and *in vitro*. The Q_{10} of the maximum isotonic power output is higher at low temperatures because of the combined influences of temperature on shortening velocity and force (Figs. 4 and 7). This analysis also suggests a mechanism for mediating the improved jumping performance seen in some anurans that breed in cold climates (John-Alder *et al.*, 1988). Little evidence exists for major adaptive shifts in the intrinsic shortening velocity, however, the thermal dependence of maximum force is variable (Bennett, 1984; Johnston and Gleeson, 1987). Maintaining a high P_0 at low temperatures could improve performance considerably (Renaud and Stevens, 1984).

I conclude that, overall, the thermal dependence of jumping is predictable on the basis of the underlying thermal dependence of maximal muscle power output. However, even after converting the R_{10} of jumping distance to the Q_{10} of power output, the thermal dependence found *in vivo* is consistently somewhat lower than that found *in vitro*. This discrepancy may be even greater at warmer temperatures. For example, *Hyla crucifer* has a very low R_{10} of jumping distance at temperatures above 20°C (John-Alder *et al.*, 1988), despite apparently having muscles with a much higher Q_{10} of shortening velocity and a stable P_0 over the same range of temperatures (John-Alder *et al.*, 1989). Whether this discrepancy results from behavioral factors or from other aspects of the *in vivo* use of muscles, for example, elastic storage of energy, remains to be determined.

V. Detailed Studies of Jumping

Surprisingly little information is available in the literature regarding the detailed kinematics or dynamics of jumping by anurans. Such information would be useful in interpreting the comparative data on jumping distance and in defining the roles of the muscles and connective tissue elements in powering the jump. Limited information is available from three types of studies: (1) measurements of ground reaction forces during takeoff using force plates; (2) measurements of the kinematics of the body based on filming; and (3) measurements of muscle activity and shortening based on EMG records, filming, and sonomicrometer recordings.

A. Ground Reaction Forces

Three studies present information based on recording ground reaction forces (Zug and Altig, 1978; Calow and Alexander, 1973; Hirano and Rome, 1984). The tambour device used by Zug and Altig (1978) did not resolve horizontal and vertical components of the force and probably lacked the necessary accuracy and speed to produce useful data. The contact times they reported in this study are only 60% of the values expected given the forces measured. Both Calow and Alexander (1973) and Hirano and Rome (1984) used force plates that resolved the vertical and horizontal components of force. Calow and Alexander's force plate measured the forces produced by a single foot and registered both the posteriorly and laterally directed components of the horizontal force. The lateral force was small and variable, reflecting variations in the takeoff direction.

Unfortunately, both of these studies using force plates report performance in jumps that were likely substantially below the maximum abilities of the frogs used. On the basis of their force plate records, Hirano and Rome (1984) calculate $L_{j,avemax}$ for *Rana pipiens* of 0.27 and 0.38 m at 18° and 25°C, respectively. The strongest jump they recorded at 25°C resulted in a calculated jumping distance of 0.47m. These values are less then half of those found by other investigators for the same species or the closely related southern leopard frog (*Rana utricularia*) (Rand, 1952; Zug, 1978; Renaud and Stevens, 1983b; R. L. Marsh *et al.*, unpublished). Calow and Alexander (1973) report that the strongest jumps undertaken by approximately 25-g *Rana temporaria* resulted in a takeoff velocity of 1.8 m s^{-1} at an angle of approximately 55°. This takeoff velocity and angle would result in a total jumping distance of about 0.47 m. However, the published force plate records do not substantiate the takeoff velocity reported. Integrating the forces shown in their Fig. 10 and doubling them to account for the force of both legs results in calculated takeoff velocities of 1.3 and 1.6 m s^{-1} at angles of about 30° for their jumps b and d, respectively. [This calculation assumes that both jumps were by a frog weighing 19 g (R. McN. Alexander, personal communication).] The low takeoff angle is expected given the relatively low force ($<1.6M_b \cdot g$) generated in the jump (Fig. 3). The resultant of the average vertical and horizontal forces was applied to the ground at an angle of over 50° for both jumps, but correcting the vertical force for the weight of the frog results in a lower predicted takeoff angle and velocity. These results suggest that the total distance jumped was less than 0.35 m for these two jumps, and data in their Table V suggest that the strength of the other jumps analyzed was similar. Unfortunately, comparative data on jumping distance are not available for *Rana temporaria*. Its overall morphology is similar to that of *Rana pipiens*, which is capable of much longer jumps, but the muscle masses reported by Calow and Alexander (1973) are smaller than those found in *Rana pipiens* (Table V). Interestingly, both Hirano and Rome (1984) and Calow and Alexander (1973) used electrical shocks to stimulate jumping, which suggests that this method may not result in maximal jumping performance.

Despite the fact that these studies report data on apparently submaximal jumps, they do provide valuable insights into the details of jumping by anurans. Calow and Alexander (1973) provide the only original records of the ground reaction forces produced by jumping frogs. After a period of variable force at the start of the jump, the ground reaction force apparently rises very quickly and is sustained for a substantial part of the contact phase. Based on the two original records provided, the average vertical and horizontal forces during the jump are 0.66 and 0.64 times the respective peak forces. Hirano and Rome (1984) provide data on the maximum resultant force during jumping at four T_b's, and the average force can be calculated from their data on vertical and horizontal takeoff velocities. In this study, as well, the average force is about 65% of the maximum force.

Data in Hirano and Rome (1984) also allow a comparison of peak and average power during jumping. Calculations of peak power are presented directly in the paper, and average power can be calculated from other data provided. Average power during these relatively short jumps is at most 47% of the peak power. If a similar ratio applies to stronger jumps, the peak power output during jumping can be impressively high. Using data from my laboratory for *Rana pipiens* jumping with a T_b of 20°C, the average power output throughout the jump would have to be about 37 W kg^{-1} of body mass to produce jumping distances equal to the $L_{j,avemax}$ (0.74 ± 0.08 m) and 85 W kg^{-1} to produce a jump equal to the $L_{j,record}$ (1.25 m). Peak power during the jump would be approximately double these values, that is, greater than 170 W kg^{-1} in the longest jump. Because the muscles that power the jump are at most 20% of the total body mass, this amounts to a peak power of more than 850 W kg^{-1} of muscle.

B. Kinematics

Several studies have reported data on jumping based on filming (Hirsch, 1931; Gans, 1961; Calow and Alexander, 1973; Altevogt *et al.*, 1986). However, only Hirsch (1931) has published detailed kinematic data that can be compared with the force plate studies. Unfortunately, the accelerations calculated by Hirsch are not consistent with the force plate records of Calow and Alexander (1973). Hirsch reports that acceleration, and thus ground reaction force, is maximal in early takeoff and then declines rapidly. Hirsch did not analyze the early low-velocity portion of the jump. Even so, his calculations show no hint of the plateau in ground reaction force that is evident in the records of Calow and Alexander (1973).

For comparison, I provide here a similar analysis of a single jump of a 31-g *Osteopilus septentrionalis*, a hylid tree frog (Marsh and John-Alder, 1994). This frog was filmed at a framing speed of 303 Hz and the position of the approximate center of mass plotted using frame-by-frame analysis (Fig. 14). The takeoff velocity in this jump was about 2.5 m s^{-1} (somewhat less than the maximum takeoff velocity of 2.8 m s^{-1} recorded in another jump by the same animal). Horizontal and vertical velocities rose slowly for the first 50 ms of the jump and then increased rapidly until takeoff at about 120 ms (Figs. 15A). Forces calculated from these data display a similar time course to those recorded by Calow and Alexander (1973). Early in the jump the forces exerted on the ground are low and then increase to a peak before falling rapidly as the frog approaches takeoff (Fig. 15B). These calculations should be taken as a rough confirmation of the pattern of force application seen by Calow and Alexander, but the filming data are subject to substantial errors early in the jump when the velocities are low. The total power required for the jump is calculated as the sum of the powers required to accelerate the frog upward and forward ($\dot{W}_j = F_v V_v + F_h V_h$) (Fig. 15C). The maximum body-mass-specific power output during this jump was about 70 W kg^{-1}. The average power during the entire jump was 33 W kg^{-1}, or 45% of the peak value.

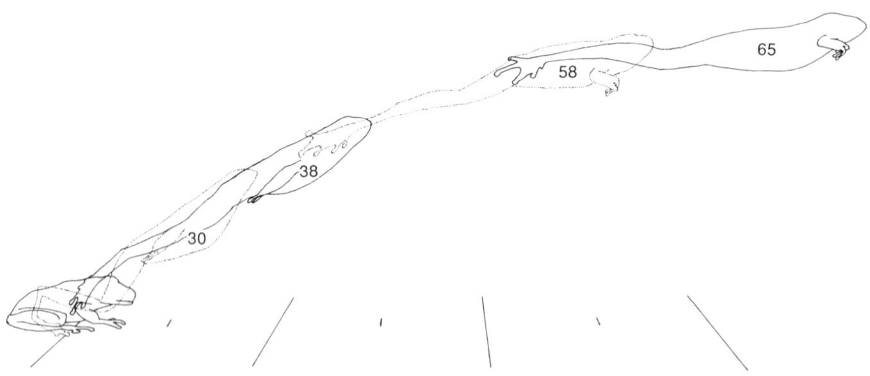

FIG. 14. Tracings from a high-speed cine film of a jump by the Cuban tree frog *Osteopilus septentrionalis*. The framing speed was 303 Hz. The numbers indicate the number of frames after the first detectable movement. (Drawn by John Kelley).

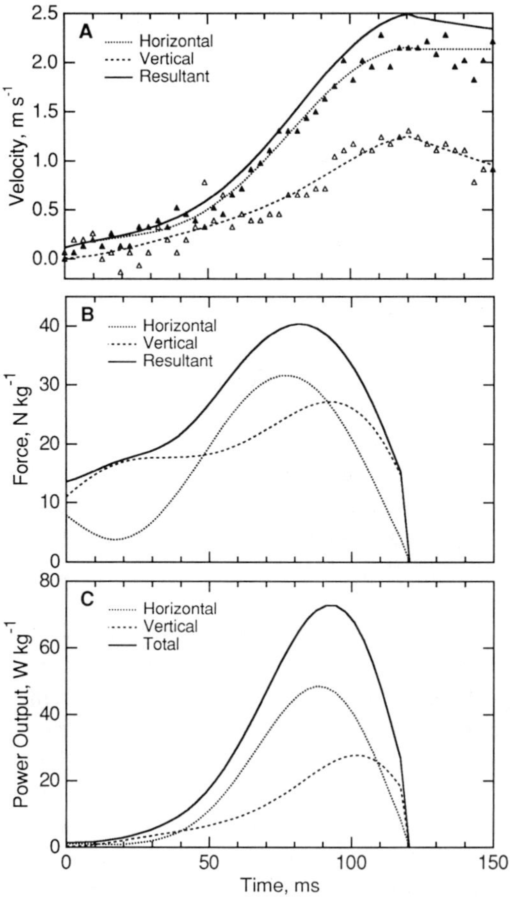

FIG. 15. (A) Velocity, (B) ground reaction force, and (C) power output during a jump by a 31-g Cuban tree frog (*Osteopilus septentrionalis*). The position of the center of mass was determined by frame-by-frame analysis. The data on x and y coordinates were fitted using an eigth-order polynomial and the smooth velocity curves were calculated from the polynomial curves. (After takeoff the horizontal velocity is assumed to remain constant and the vertical velocity declines as determined by the acceleration of gravity.) The individual points plotted for vertical and horizontal velocities are 3-point running averages of values determined from the raw position data. Force and power are calculated from the smoothed velocity curves and expressed per unit body mass. Reproduced with permission from Marsh and John-Alder, 1994.

C. Muscle Activity and Shortening

The impressive mechanical performance of frogs during jumping implies that the complex set of muscles that propels the frog shortens under very favorable conditions. Predicting precisely what these conditions should be and testing these predictions is difficult. As a first approximation, however, the following predictions can be made. First, the muscles should be active and shorten throughout most of the jump. Second, the velocity of shortening should be within the range that provides high power output (Fig. 5). Lengthening has been reported to enhance performance of some skeletal muscles (Woledge *et al.*, 1985), but major lengthening of active muscles would actually be expected to reduce the average power output of the hindlimb muscles. Anuran jumping does not involve a countermovement before the jump, and energy to stretch active muscles must come from the contraction of other muscles. A short period of lengthening or isometric contraction early in the jump might enhance performance because force would build up before shortening begins.

Calow and Alexander (1973) examined muscle shortening based on data from high-speed films and combined these data with estimated forces based on their force plate records in an attempt to gain a complete picture of the *in vivo* function of the major muscles. They predicted the fractional shortening and shortening velocity of the muscle fibers in six muscles using data on joint angle from the films along with measurements of the moment arms of the muscles as a function of joint angle. They estimated that the various muscles show substantial differences in fractional shortening and shortening velocity estimated in the later stages of takeoff (Table IX). The largest strains amounted to 30–50% of prejump resting length (L_r). This level of strain was calculated for three pinnate muscles, the curalis, the plantaris longus, and the semitendinosus. In contrast, the three parallel-fibered muscles had much smaller strains (9–15% of L_r). Shortening velocity was also estimated to be higher in the pinnate muscles (>3.7 $L_r\ s^{-1}$) compared with the parallel-fibered muscles ($<1.4\ L_r\ s^{-1}$). These calculations assume that all the shortening occurred in the muscle fibers and none in tendons or other connective tissue elements.

Calow and Alexander (1973) calculated the forces acting on the muscles in the later stages of takeoff based on their force plate records and measures of the moment arms of the muscles. They use the ground reaction forces to calculate the moments about the ankle, knee, and hip, and from these moments they calculate the stress (force/area) in the muscles. The calculation is straightforward for the plantaris, be-

TABLE IX

MUSCLE SHORTENING AND STRESS (FORCE PER CROSS-SECTIONAL AREA) CALCULATED BY CALOW AND ALEXANDER (1973) FROM FORCE PLATE RECORDS AND HIGH-SPEED FILMS OF *RANA TEMPORARIA* JUMPING AT 23°C[a]

Muscle	Strain (L_0^{-1})	Strain rate		Stress	
		L_0 s^{-1}	$V V_{max}^{-1}$	kN m^{-2}	$P P_0^{-1}$
Plantaris longus	0.29	5.0	0.62	110	0.36
Curalis	0.37	4.8	0.60	100	0.33
Gluteus magnus	0.10	1.4	0.18	100	0.33
Semitendinosus	0.50	3.7	0.46	65	0.21
Gracilis major	0.15	1.2	0.15	65	0.21
Semimembranosus	0.09	0.6	0.07	65	0.21

[a] $V V_{max}^{-1}$ calculated assuming a V_{max} of $8L_0$ s^{-1} (see Table III).

cause it is the only muscle acting to extend the ankle. However, for the muscles of the thigh, an analytical solution is not feasible without some simplifying assumptions. They simplify the situation by assuming that the groups of thigh muscles, the extensors of the knee (curalis and gluteus magnus), and the extensors of the hip (semitendinosus, semimembranosus, and gracilis major) have similar stresses within each group. Their results indicate stresses of about 100 kN m^{-2} for the plantaris and the knee extensors and only 65 kN m^{-2} for the hip extensors.

Unfortunately, the values calculated by Calow and Alexander (1973) for stress and strain are not consistent with one another. Although we do not know if muscles operating *in vivo* follow precisely their *in vitro* force–velocity curves, the stress produced by a muscle is expected to be inversely proportional to its velocity of shortening. The inconsistency between the predicted velocity and force values might be improved by dispensing with the assumption of uniform stress within each of the muscle groups in the thigh. However, the calculated shortening velocity of the curalis (~5.0 L_r s^{-1}) is still too high to allow this muscle to exert the required force. A similar discrepancy exists for the plantaris, which is predicted to have a similarly high shortening velocity. Assuming a V_{max} at 22°C of 8 L_0 s^{-1}, a typical anuran force–velocity curve (Table III), and L_r approximately equal to L_0, these muscles should produce a force of only $0.12P_0$, or about 60 kN m^{-2} when shortening at 5 L_r s^{-1}. The inconsistencies in the data may be explained by methodological errors, but another potential source of error is the assumption that all the shortening is due to the muscle fibers. Calow and Alexander calculated the largest shortening velocities for the pinnate muscles that also have the longest tendons. If these tendons were stretched early in takeoff, elastic recoil later in the jump would reduce the shortening velocity to be accounted for by the muscle fibers. Additionally, other thigh muscles may also be involved in jumping, for example, the adductor magnus (Section III, B).

Two other groups of investigators have attempted to measure the shortening velocities of the muscles used in jumping. On the basis of filming studies of *Rana pipiens*, Lutz and Rome (1994) report that the semimembranosus shortens by about 22% during jumping. This strain is greater than that reported by Calow and Alexander (1973) for the same muscle. The velocity of shortening of this muscle was found to be constant for most of the jump at 3.4 L_0 s^{-1}. This value is within those predicted to produce nearly maximum power in isovelocity contractions.

Olson and Marsh (1992) measured EMG activity and shortening in the semimembranosus and the gluteus magnus muscle of jumping *Rana catesbeiana*. In these large frogs (mean mass, 485 g), shortening of the muscle fibers was measured directly using sonomicrometer transducers implanted parallel to a segment of the fibers (Griffiths, 1987, 1991; Marsh *et al.*, 1992). The pattern of shortening and activation seen in the semimembranosus of *Rana catesbeiana* during jumping is consistent with the hypothesis that this muscle produces power for most of the takeoff phase. Shortening occupies most of the takeoff phase and follows a consistent trajectory during most jumps (Fig. 16). Electromyographic recordings indicate that the semimembranosus is activated just before the beginning of shortening. The time from the beginning of the EMG activity to the beginning of shortening averages 25 ms and is independent of jumping distance. This brief delay before shortening should allow some buildup of force prior to shortening. In some cases the semimembranosus lengthens just prior to shortening (Fig. 16), but this lengthening is usually less than 2% of L_r (mean, 0.97 $\pm$ 0.04% L_r). The strain in this muscle is dependent on jumping distance (Fig. 17), increasing from about 16% of the prejump resting length (L_r) at a jumping distance of 0.4 m to more than 25% of L_r when the frogs jumped 1 m. The average velocity of shortening during takeoff also varies with jumping distance. As jumping distance increases from 0.4 to 1.0 m, the shortening velocity of the semimembranosus almost doubles, going from 0.9 to 1.7 L_r s^{-1} (Fig. 18). If L_r is again assumed to be close to L_0, these velocities correspond to 0.15 to 0.28 V V_{max} $^{-1}$. Interpreting these shortening velocities requires some cau-

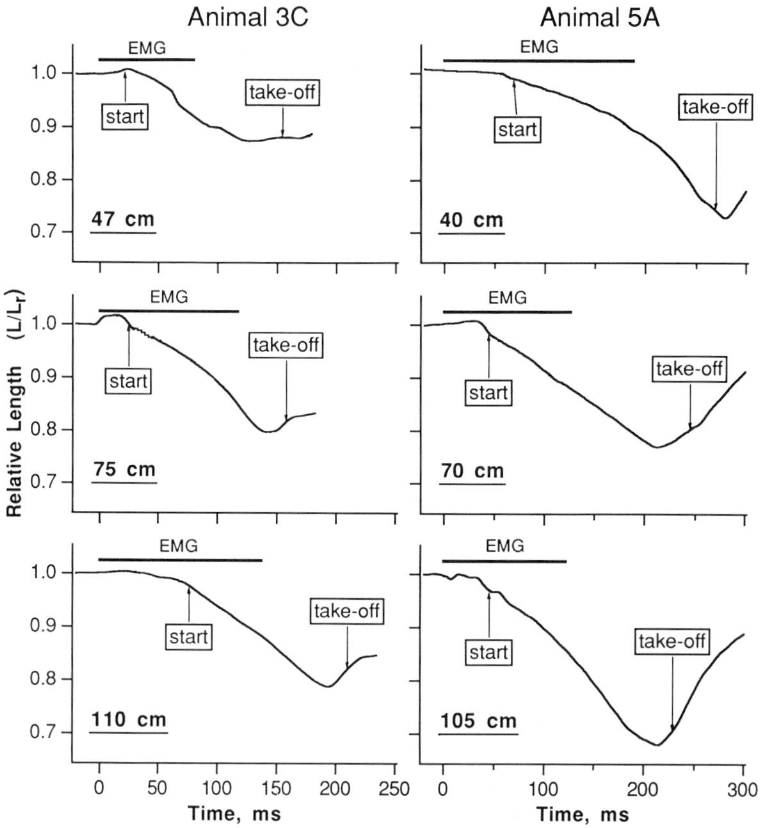

FIG. 16. Length during jumping of a segment of the semimembranosus muscle of *Rana catesbeiana* as determined by implanted sonomicrometer transducers. Length is expressed relative to the length just prior to the jump (L_r). The times of the start of the jump and takeoff were determined by simultaneous video recordings at 60 Hz. The solid black bar above each trace gives the duration of muscle activity as determined by simultaneous electromyographic recordings. Jump length is indicated in each panel.

tion, however, as they represent the average velocities during shortening and the velocity during a jump can vary considerably (Fig. 16). The implications of this variation for the production of mechanical power need to be investigated. Also the muscle is deactivated as the jump proceeds and the production of power might be expected to decline in the later stages of takeoff. However, the EMG activity does not reflect completely the period of time when the muscle is active and producing force. A short delay, 5–10 ms, is expected after the muscle is first stimulated, and after stimulation ceases the muscle remains active for

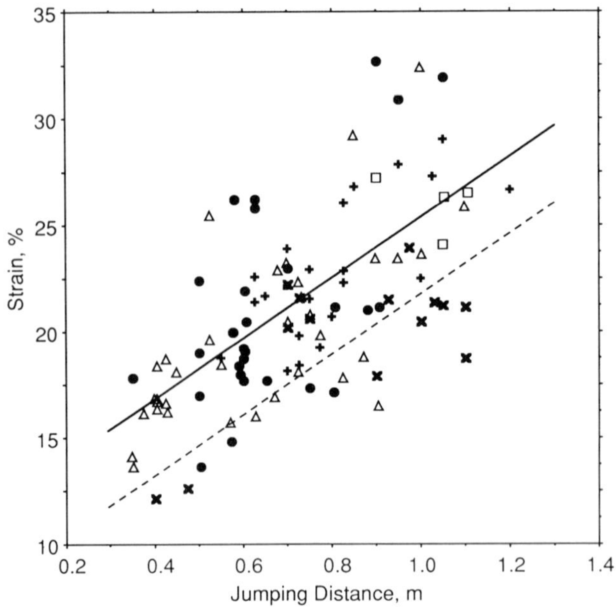

FIG. 17. Strain during jumping of a segment of the semimembranosus muscle of *Rana catesbeiana* as determined by implanted sonomicrometer transducers. Among four animals the effect of jumping distance was statistically indistinguishable (ANCOVA), and the solid curve is the regression equation that describes those data. A single animal indicated by the symbol (×) had lower strains and the dashed line applies to that animal.

a period of time (approximately 60 ms based on the twitch kinetics of bullfrog muscle; Table III).

In contrast to the semimembranosus, the gluteus magnus is active under conditions that would appear to be suboptimal for power output. This muscle displays a variable length trajectory, particularly when different animals are compared, and in general has lower strains than the semimembranosus (Fig. 19). The variation in the pattern of shortening of the gluteus magnus is apparently due to variation in the rates of extension of the hip and knee, and the presence of substantial moment arms at both joints. The total strain of the gluteus magnus (including any lengthening early in the jump) varies greatly among individual animals, ranging from 12 to 21% of L_r, and is not significantly correlated with jumping distance. In many cases this muscle remained isometric or lengthened for one-half or more of the takeoff phase (Fig. 19).

VI. Conclusions

Anuran jumping provides the opportunity to examine the relations between jumping performance and the properties of the muscles that power this movement. The following general conclusions may be drawn from this analysis.

1. Jumping performance is accurately predicted from simple ballistics formulae if the distance traveled prior to takeoff and the takeoff height are considered in the analysis.
2. The effects of body size and temperature on maximal jumping ability appear to be predictable consequences of the effects of these variables on the mechanics of jumping and the ability of the hindlimb muscles to produce mechanical power.
3. The average power output during jumping by many species can equal or exceed the maximal power available from the muscles as

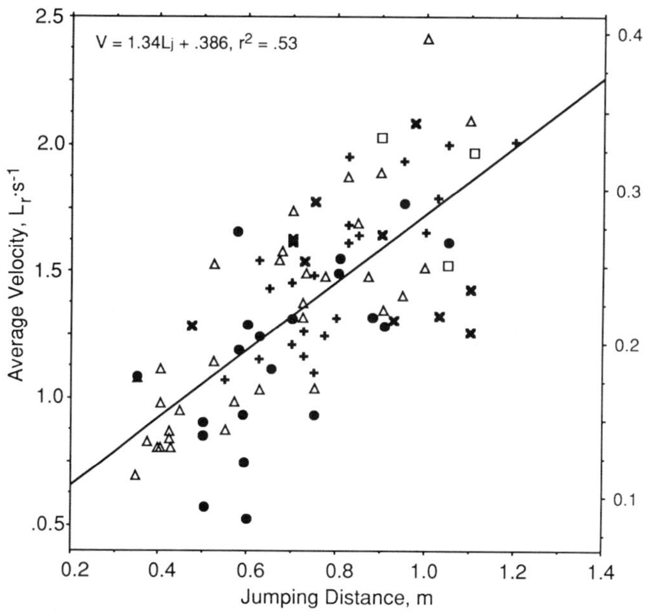

FIG. 18. Velocity of shortening during jumping of a segment of the semimembranosus muscle of *Rana catesbeiana* as determined by implanted sonomicrometer transducers. The V_{max} used to calculate the relative velocity scale on the right-hand axis was determined from the allometric relation for the sartorius muscle *Rana catesbeiana* (Fig. 6).

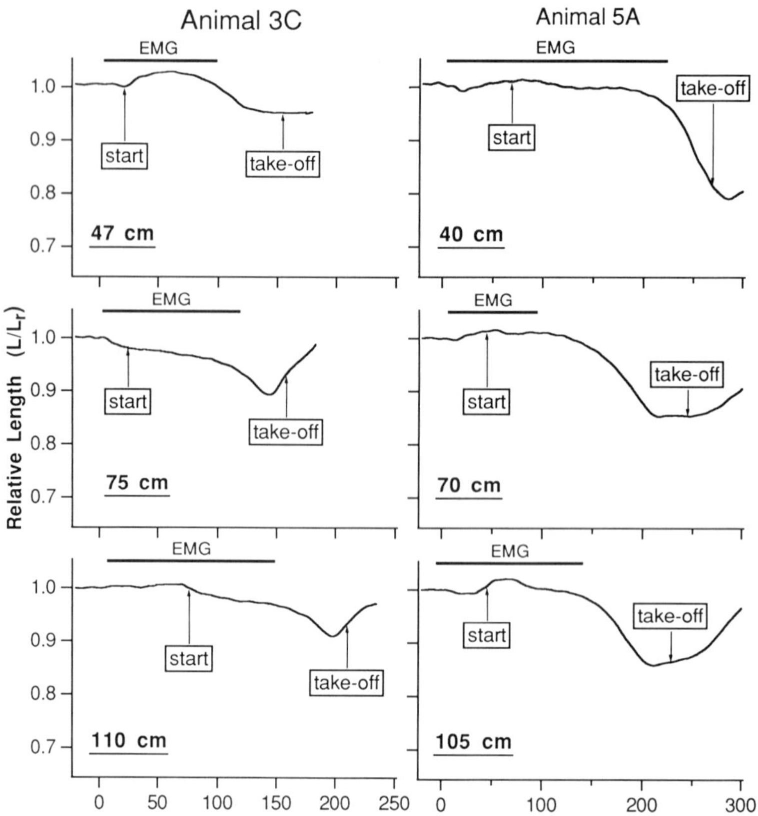

FIG. 19. Length during jumping of a segment of the gluteus magnus muscle of *Rana catesbeiana* as determined by implanted sonomicrometer transducers. Length is expressed relative to the length just prior to the jump (L_r). The times of the start of the jump and takeoff were determined by simultaneous video recordings at 60 Hz. The solid black bar above each trace gives the duration of muscle activity as determined by simultaneous electromyographic recordings. Jump length is indicated in each panel.

predicted from *in vitro* isotonic contractions. This equivalence implies that many of the hindlimb muscles must operate under conditions that allow maximal power output throughout most of takeoff.

4. The maximal power required during takeoff is approximately twice the average value and thus greatly exceeds the predicted ability of the muscles to produce power based on isotonic contractions. Elastic storage of energy could redistribute the power output during takeoff and account for the apparent discrepancy between the power produced in jumping and that available from the muscles.

Future research should be directed toward identifying the structures that store and release energy during jumping and examining the mechanical conditions under which the major muscles operate. With data on strain, shortening velocity, and EMG activity, it may be possible to simulate the mechanical conditions prevailing *in vivo* (Marsh *et al.*, 1992; Lutz and Rome, 1994) and measure power output in a more realistic manner than using isotonic contractions.

VII. Abbreviations

A_m	Cross-sectional area of muscle fibers
ANCOVA	Analysis of covariance
EMG	Electromyogram
F_t	Average ground reaction force during takeoff
g	Acceleration due to gravity
H_a	Maximum height of the center of mass during the aerial phase of the jump
H_t	Height of the center of mass at takeoff
L_{cm}	Length along the outstretched hindlimbs from the tips of the toes to the center of mass
L_{hl}	Length of the outstretched hindlimbs
L_j	Total jumping distance
$L_{j,ave}$	Mean jumping distance based on all the jumps recorded in a trial.
$L_{j,avemax}$	Mean maximal jumping distance, calculated as the mean of the longest jumps recorded for each individual in a trial
$L_{j,d}$	Horizontal distance traveled after the frog descends below H_t
$L_{j,l}$	Distance traveled by the center of mass from takeoff until the center of mass returns to H_t
$L_{j,record}$	The longest jump recorded for any individual in a trial
$L_{j,t}$	Horizontal distance traveled by the center of mass before takeoff
L_r	Prejump resting length of a muscle
L_{sv}	Snout–vent length
L_0	Muscle length at which isometric force is maximal
M_b	Body mass
M_m	Muscle mass
P_0	Maximal active force in a fully fused isometric tetanus

Q_{10}	Thermal dependence of a rate function expressed as the fractional change for a 10°C increase in temperature
R_{10}	Thermal dependence of a variable which is not a rate function expressed as the fractional change for a 10°C increase in temperature
$R_{\dot{W}}$	Power ratio
$t_{a,l}$	Time required to travel $L_{j,l}$
T_b	Body temperature
t_c	Contact time during takeoff
t_f	Time required to fall from H_a to the ground
t_r	Time required to rise from H_t to H_a
T_M	Muscle temperature
tP_{tw}	Time to peak force in a twitch
$t50\%R$	Time from peak force to 50% relaxation in a twitch
V_v	Vertical velocity of the center of mass at takeoff
V_h	Horizontal velocity of the center of mass at takeoff
V_{max}	Maximal velocity of shortening determined by extrapolating the force velocity curve to zero force
V_t	Takeoff velocity (resultant of V_h and V_v)
W_K	Kinetic energy at takeoff
W_P	Potential energy at takeoff
W_{tot}	Total energy at takeoff
$\dot{W}_j$	Total mechanical power expended during takeoff
$\dot{W}_{j,k}$	Power required to accelerate the frog to V_t
$\dot{W}_{max}$	Maximal power measured during isotonic contractions
θ	Takeoff angle

Acknowledgments

The preparation of this review and the new work reported were supported by NIH Grant AR39318. Mr. V. Wallach kindly tracked down information on *Litoria nasuta*. Dr. R. McN. Alexander graciously responded to my request for more information regarding the force plate measurements in Calow and Alexander (1973). Ms. S. Sarbadhikary and Mr. S. Massoni assisted in measuring the maximal jumping ability of *Rana pipiens*. I am grateful to Dr. H. A. John-Alder for stimulating my interest in anuran jumping. Dr. J. M. Olson contributed extensively to the work on bullfrogs reported here. Dr. R. J. Wasserzug provided many useful comments on the manuscript.

References

Alexander, R. McN., Jayes, A. S., Maloiy, G. M. O., and Wathuta, E. M. (1981). Allometry of leg muscles in mammals. *J. Zool.*, **194,** 539–552.

Altevogt, R., Holtmann, H., and Kaschek, N. (1986). High frequency cinematography studies on locomotion and preying in Indian skitter frogs *Rana cyanophlyctis* Schneider. *J. Bombay Nat. Hist. Soc.* **83,** Suppl., 102–111.

Anderson, B. D., Feder, M. E., and Full, R. J. (1991). Consequences of a gait change during locomotion in toads (*Bufo woodhousii fowleri*). *J. Exp. Biol.* **158,** 133–148.

Beiser, A. (1978). "Physics," 2nd ed., Benjamin/Cummings, Menlo Park, CA.

Bennet-Clark, H. C. (1977). Scale effects in jumping animals. *In* "Scale Effects in Animal Locomotion" (Pedley, T. J., ed.), pp. 185–201. Academic Press, New York.

Bennett, A. F. (1984). Thermal dependence of muscle function. *Am. J. Physiol.* **247,** 217–229.

Bennett, A. F. (1990). The thermal dependence of locomotor capacity. *Am. J. Physiol.* **259,** R253–R258.

Burchfield, D. M., and Rall, J. A. (1986). Temperature dependence of the crossbridge cycle during unloaded shortening and maximum isometric tetanus in frog skeletal muscle. *J. Muscle Res. Cell Motil.* **7,** 320–326.

Calow, L. J., and Alexander, R. McN. (1973). A mechanical analysis of the hind leg of a frog (*Rana temporaria*). *J. Zool.* **171,** 293–321.

Cecchi, G., Columo, F., and Lombardi, V. (1978). Force–velocity relation in normal and nitrate-treated frog single muscle fibers during the rise of tension in isometric tetanus. *J. Physiol. (London)* **285,** 257–273.

Close, R. (1965). Force: Velocity properties of mouse muscles. *Nature (London)* **206,** 718–719.

Close, R. I. (1972). Dynamic properties of mammalian muscles. *Physiol. Rev.* **52,** 129–197.

Dobrowolska, H. (1973). Body-part proportions in relation to the mode of locomotion in anurans. *Zool. Pol.* **23,** 59–108.

Duellman, W. E., and Trueb, L. (1986). "Biology of Amphibians." McGraw-Hill, New York.

Dunlap, D. G. (1960). The comparative myology of the pelvic appendage in the Salientia. *J. Morphol.* **106,** 1–76.

Edman, K. A. P. (1979). The velocity of unloaded shortening and its relation to sarcomere length and isometric force in vertebrate muscle fibers. *J. Physiol. (London)* **291,** 143–159.

Edman, K. A. P., and Hwang, J. C. (1977). The force–velocity relationship in vertebrate muscle fibres at varied tonicity of the extracellular medium. *J. Physiol. (London)* **269,** 255–272.

Edman, K. A. P., Mulieri, L. A., and Scubon-Mulieri, B. (1976). Non-hyperbolic force–velocity relationship in single muscle fibers. *Acta Physiol. Scand.* **98,** 143–156.

Else, P. L., and Bennett, A. F. (1987). The thermal dependence of locomotor performance and muscle contractile function in the salamander *Ambystoma tigrinum nebulosum*. *J. Exp. Biol.* **128,** 219–233.

Emerson, S. B. (1978). Allometry and jumping in frogs: Helping the twain to meet. *Evolution (Lawrence, Kans.)* **32,** 551–564.

Emerson, S. B. (1985). Jumping and leaping. *In* "Functional Vertebrate Morphology" (Hildebrand, M. E., Bramble, D. M., Liem, K. F. and Wake, D. B., eds.), pp. 58–72. Harvard Univ. Press, Cambridge, MA.

Emerson, S. B., and De Jongh, H. J. (1980). Muscle activity at the ilio-sacral articulation of frogs. *J. Morphol.* **166,** 129–144.

Fenn, W. O., and Marsh, B. S. (1935). Muscular force at different speeds of shortening. *J. Physiol. (London)* **85,** 277–297.

Gans, C. (1961). A bullfrog and its prey: A look at the biomechanics of jumping. *Nat. Hist. (N.Y.)* **70,** 26–37.

Gans, C., and de Gueldre, G. (1992). Striated muscle: Physiology and functional morphology. *In* "Environmental Physiology of the Amphibians" (Feder, M. E. and Burggren, W. W., eds.), pp. 277–313. Univ. of Chicago Press, Chicago.

Garland, T., Jr.. (1984). Physiological correlates of locomotory performance in a lizard: An allometric approach. *Am J. Physiol.* **247,** R806–R815.

Garland, T., Jr. (1985). Ontogenetic and individual variation in size, shape and speed in the Australian agamid lizard *Amphibolurus nuchalis. J. Zool.* **207,** 425–439.

Gatten, R. E., Jr., Miller, K., and Full, R. J. (1992). Energetics at rest and during locomotion. *In* "Environmental Physiology of the Amphibians" (Feder, M. E. and Burggren, W. W., eds.), pp. 314–377. Univ. of Chicago Press, Chicago.

Gibbs, C. L., and Chapman, J. B. (1974). Effects of stimulus conditions, temperature, and length on energy output of frog and toad sartorius. *Am. J. Physiol.* **227,** 964–971.

Gray, J. (1968). "Animal Locomotion." Weidenfeld & Nicolson, London.

Griffiths, R. I. (1987). Ultrasound transit time gives direct measurement of muscle fibre length *in vivo. J. Neurosci. Methods* **21,** 159–165.

Griffiths, R. I. (1991). Shortening of muscle fibres during stretch of the active cat medial gastrocnemius muscle—The role of tendon compliance. *J. Physiol. (London)* **436,** 219–236.

Hill, A. V. (1922). The maximum work and mechanical efficiency of human muscles, and their most economical speed. *J. Physiol. (London)* **56,** 19–41.

Hill, A. V. (1938). The heat of shortening and the dynamic constants of muscle. *Proc. R. Soc. London, Ser. B* **B 126,** 136–195.

Hill, A. V. (1950). The dimensions of animals and their muscular dynamics. *Sci. Prog. (Oxford)* **38,** 209–230.

Hill, A. V. (1965). "Trials and Trails in Physiology." Edward Arnold, London.

Hill, A. V. (1970). "First and Last Experiments in Muscle Mechanics." Cambridge Univ. Press, Cambridge, UK.

Hill, A. V., and Hartree, W. (1920). The four phases of heat-production of muscle. *J. Physiol. (London)* **54,** 84–128.

Hirano, M., and Rome, L. C. (1984). Jumping performance of frogs (*Rana pipiens*) as a function of muscle temperature. *J. Exp. Biol.* **108,** 429–439.

Hirsch, W. (1931). Zur physiologischen Mechanik der Froschsprunges. *Z. Vergl. Physiol.* **15,** 1–49.

Huey, R. B., and Stevenson, R. D. (1979). Integrating thermal physiology and ecology of ectotherms: a discussion of approaches. *Amer. Zool.* **19,** 357–366.

Iaizzo, P. A. (1990). Histochemical and physiological properties of *Rana temporaria tibialis* anterior and lumbricalis IV muscle fibres. *J. Muscle Res. Cell Motil.* **11,** 281–292.

John-Alder, H. B., and Morin, P. J. (1990). Effects of larval density on jumping ability and stamina in newly metamorphosed *Bufo woodhousii fowleri. Copeia*, 1990, 856–860.

John-Alder, H. B., Morin, P. J., and Lawler, S. (1988). Thermal physiology, phenology, and distribution of tree frogs. *Amer. Nat.* **132,** 506–520.

John-Alder, H. B., Barnhart, M. C., and Bennett, A. F. (1989). Thermal sensitivity of swimming performance and muscle contraction in northern and southern populations of tree frogs (*Hyla crucifer*). *J. Exp. Biol.* **142,** 357–372.

Johnston, I. A., and Gleeson, T. T. (1987). Effects of temperature on contractile properties of skinned muscle fibers from three toad species. *Am. J. Physiol.* **252,** R371–R375.

Katz, B. (1939). The relation between force and speed in muscular contraction. *J. Physiol. (London)* **96,** 45–64.
Knowles, T. W., and Weigl, P. D. (1990). Thermal dependence of anuran burst locomotor performance. *Copeia*, 1990, 796–802.
Kreighbaum, E., and Barthels, K. M. (1990). "Biomechanics: A Quantitative Approach for Studying Human Movement." Macmillan, New York.
Lannergren, J., Lindblom, P., and Johansson, B. (1982). Contractile properties of two varieties of twitch fibers in *Xenopus laevis*. *Acta Physiol. Scand.* **114,** 523–535.
Londos, P. L., and Brooks, R. J. (1988). Effect of temperature acclimation on locomotory performance curves in the toad, *Bufo woodhousii woodhousii*. *Copeia*, 1988, 26–32.
Londos, P. L., and Brooks, R. J. (1990). Time course of temperature acclimation of locomotory performance in the toad, *Bufo woodhousei woodhousei*. *Copeia*, 1990. 827–835.
Luff, A. R. (1981). Dynamic properties of the inferior rectus, extensor digitorum longus, diaphragm and soleus muscles of the mouse. *J. Physiol. (London)* **313,** 161–171.
Luff, A. R., and Proske, U. (1976). Properties of motor units of the frog sartorius muscle. *J. Physiol. (London)* **258,** 673–685.
Luff, A. R., and Proske, U. (1979). Properties of motor units of the frog iliofibularis muscle. *Am. J. Physiol.* **236,** C35–C40.
Lutz, G. J., and Rome, L. C. (1994). Built for jumping: The design of the frog muscular system. *Science* **263,** 370–372.
Mai, M. T., and Lieber, R. L. (1990). A model of semitendinosus muscle sarcomere length, knee and hip joint interaction in the frog hindlimb. *J. Biomech.* **23,** 271–279.
Marsh, R. L. (1988). Ontogenesis of contractile properties of skeletal muscle and sprint performance in the lizard *Dipsosaurus dorsalis*. *J. Exp. Biol.* **137,** 119–139.
Marsh, R. L. (1990). Deactivation rate and shortening velocity as determinants of contractile frequency. *Am. J. Physiol.* **259,** R223–R230.
Marsh, R. L., and Bennett, A. F. (1985). Thermal dependence of isotonic contractile properties of skeletal muscle and sprint performance of the lizard *Dipsosaurus dorsalis*. *J. Comp. Physiol.* **155,** 541–551.
Marsh, R. L., and Bennett, A. F. (1986a). Thermal dependence of isotonic contractile properties of skeletal muscle from the lizard *Sceloporus occidentalis* with comments on methods for fitting and comparing force–velocity curves. *J. Exp. Biol.* **126,** 63–77.
Marsh, R. L., and Bennett, A. F. (1986b). Thermal dependence of sprint performance of the lizard *Sceloporus occidentalis*. *J. Exp. Biol.* **126,** 79–87.
Marsh, R. L., and John-Alder, H. A. (1994). Jumping performance of hylid frogs measured with high-speed cine film. *J. Exp. Biol.* **188,** 131–141.
Marsh, R. L., and Taigen, T. L. (1987). Properties enhancing aerobic capacity of calling muscles in gray tree frogs *Hyla versicolor*. *Am. J. Physiol.* **252,** R786–R793.
Marsh, R. L., Olson, J. M., and Guzik, S. K. (1992). Mechanical performance of scallop adductor muscle during swimming. *Nature (London)* **357,** 411–413.
Matsui, M. (1978). Correlation between relative weights of hindlimb muscles and locomotor patterns in Japanese anurans. *Contrib. Biol. Lab. Kyoto Univ.* **25,** 223–240.
McMahon, T. A. (1984). "Muscles, Reflexes, and Locomotion." Princeton Univ. Press, Princeton, NJ.
Mendiola, P., Decosta, J., Lozano, M. T., and Agulleiro, B. (1991). Histochemical determination of muscle fiber types in locomotor muscles of anuran amphibians. *Comp. Biochem. Physiol.* **99A,** 365–369.
Olson, J. M., and Marsh, R. L. (1992). Activation and length changes in two hindlimb muscles of the frog during jumping. *Amer. Zool.* **32,** 41A.

Osgood, P. F., and Brewster, W. R. (1963). Effect of temperature and continuous stimulation on gastrocnemius of the frog. *Am. J. Physiol.* **205,** 1299–1303.

Pough, F. H., Magnusson, W. E., Ryan, M. J., Wells, K. D., and Taigen, T. L. (1992). Behavioral energetics. *In* "Environmental Physiology of the Amphibians" (Feder, M. E. and Burggren, W. W., eds.), pp. 378–436. Univ. of Chicago Press, Chicago.

Putnam, R. W. (1979). The basis for differences in lactic acid content after activity in different species of anuran amphibians. *Physiol. Zool.* **52,** 509–519.

Putnam, R. W., and Bennett, A. F. (1983). Histochemical, enzymatic, and contractile properties of skeletal muscles of three anuran amphibians. *Am. J. Physiol.* **244,** R558–R567.

Rand, A. S. (1952). Jumping ability of certain anurans, with notes on endurance. *Copeia,* 1952, 15–20.

Rand, A. S., and Rand, P. J. (1966). The relation of size and distance jumped in *Bufo marinus. Herpetologia* **22,** 206–209.

Renaud, J. M., and Stevens, E. D. (1981a). Effect of acclimation temperature and pH on contraction of frog sartorius muscle. *Am. J. Physiol.* **240,** R301–R309.

Renaud, J. M., and Stevens, E. D. (1981b). The interactive effects of temperature and pH on the isometric contraction of toad sartorius muscle. *J. Comp. Physiol.* **145,** 67–71.

Renaud, J. M., and Stevens, E. D. (1983a). A comparison between field habits and contractile performance of frog and toad sartorius muscle. *J. Comp. Physiol.* **151,** 127–131.

Renaud, J. M., and Stevens, E. D. (1983b). The extent of long-term temperature compensation for jumping distance in the frog, *Rana pipiens*, and the toad, *Bufo americanus. Can. J. Zool.* **61,** 1284–1287.

Renaud, J. M., and Stevens, E. D. (1984). The extent of short-term and long-term compensation to temperature shown by frog and toad sartorius muscle. *J. Exp. Biol.* **108,** 57–75.

Rome, L. C. (1983). The effect of long-term exposure to different temperatures on the mechanical performance of frog muscle. *Physiol. Zool.* **56,** 33–40.

Rome, L. C., Stevens, E. D., and John-Alder, H. B. (1992). The influence of temperature and thermal acclimation on physiological function. *In* "Environmental Physiology of the Amphibians" (Feder, M. E. and Burggren, W. W., eds.), pp. 183–205. Univ. of Chicago Press, Chicago.

Seow, C. Y., and Ford, L. E. (1991). Shortening velocity and power output of skinned muscle fibers from mammals having a 25,000-fold range of body mass. *J. Gen. Physiol.* **97,** 541–560.

Sperry, D. G. (1981). Fiber type composition and postmetamorphic growth of anuran hind limb muscles. *J. Morphol.* **170,** 321–345.

Stevens, E. D. (1988). Effect of pH and stimulus phase on the work done by isolated frog sartorius muscle during cyclical contraction. *J. Muscle Res. Cell Motil.* **9,** 329–333.

Taigen, T. L., Wells, K. D., and Marsh, R. L. (1985). The enzymatic basis of high metabolic rates in calling frogs. *Physiol. Zool.* **58,** 719–726.

Tracy, C. R. (1979). Further thoughts on anuran thermoregulation: Discussion. *In* "The Behavioral Significance of Color" (Burtt, E. H., ed.), pp. 63–68. Garland Publ., New York.

Tricker, R. A. R., and Tricker, B. J. K. (1967). "The Science of Movement." Am. Elsevier, New York.

Tyler, M. J., Smith, L. A., and Johnstone, R. E. (1984). "Frogs of Western Australia." Western Australian Museum, Perth.

Wermel, J. (1934). Über die Körperproportionen der Wirbeltie und ihre funktionelle

Bedeutung. Biometrische übungen Zweiter Aufsatz. Die Extremitätenproportionen und der Sprung den Salientia. *Z. Anat. Entwicklungsgesch.* **103,** 645–659.

Whitehead, P. J., Puckridge, J. T., Leigh, C. M., and Seymour, R. S. (1989). Effect of temperature on jump performance of the frog *Limnodynastes tasmaniensis. Physiol. Zool.* **62,** 937–949.

Wine, R. N., and Gatten, R. E. (1992). Glycolytic support for locomotion in *Anolis carolinensis* and *Rana pipiens*: simultaneous effects of temperature and duration of exercise. *Physiol. Zool.* **65,** 724–741.

Woledge, R. C. (1968). The energetics of tortoise muscle. *J. Physiol.* (*London*) **197,** 685–707.

Woledge, R. C., Curtin, N. A., and Homsher, E. (1985). "Energetic Aspects of Muscle Contraction." Academic Press, New York.

Zug, G. R. (1972). Anuran locomotion: Structure and function. I. Preliminary observations on the relation between jumping and the osteometrics of appendicular and postaxial skeleton. *Copeia*, 1972, 613–624.

Zug, G. R. (1978). Anuran locomotion—Structure and function. 2: Jumping performance of semiaquatic, terrestrial, and arboreal frogs. *Smithson. Contrib. Zool.* **276,** 1–31.

Zug, G. R. (1985). Anuran locomotion: fatigue and jumping performance. *Herpetologica* **41,** 188–194.

Zug, G. R., and Altig, R. (1978). Anuran jumping—Structure and function: The jumping forces of frogs. *J. Wash. Acad. Sci.* **68,** 144–147.

Exercise Performance of Reptiles

ALBERT F. BENNETT

Department of Ecology and Evolutionary Biology, University of California, Irvine, Irvine, California 92717

I. Introduction
II. Endurance Activity
  A. Cost of Locomotor Activity
  B. Endurance and Maximal Oxygen Consumption
  C. Ecological and Evolutionary Implications of Endurance
III. Burst and Exertion Activity
  A. Anaerobic Glycolysis and Metabolic Support
  B. Anaerobic Metabolism under Natural Conditions
  C. Ecological and Evolutionary Consequences
IV. Summary
  References

I. Introduction

The first quantitative measurements on reptilian exercise physiology were undertaken only three decades ago. This study (Bartholomew and Tucker, 1963) described the effect of body temperature on maximal oxygen consumption and aerobic scope in an agamid lizard, *Amphibolurus barbatus*. Since that time, reptiles have proved to be highly versatile and useful animals in which to examine activity and activity energetics, and hundreds of papers have subsequently appeared on these topics. In many areas of comparative exercise physiology (e.g., anaerobic energetics, burst speed performance, measurement of natural selection on performance), reptiles are the group in which new techniques were pioneered and remain today the best understood and most widely examined taxon. What accounts for the great popularity of reptiles for these types of studies? The group offers a unique combination of factors, some historical, some practical, including:

1. Position in the evolutionary development of the vertebrates;
2. Availability of well-understood phylogenetic relationships;
3. Diversity of locomotor morphology;
4. Diversity of ecology, especially in regard to locomotion;
5. Accessibility of natural populations.

Similar factors have also been significant in the prominence of one group of reptiles, lizards, in a wide variety of ecological studies (e.g., Milstead, 1967; Huey *et al.*, 1983).

Crucial factors in the popularity of reptiles have been both their position in the evolutionary development of the vertebrates and our extensive knowledge of the taxonomy and phylogeny of the group. ["Reptiles" in this article is used in the common, not cladistic (e.g., see Gauthier *et al.*, 1988b), sense and refers to squamates, turtles, sphenodons, crocodilians, and various extinct groups of dinosaurs and their allies.] Reptiles were the first fully terrestrial vertebrate group, the first to meet all the challenges of land existence head on, including its very different characteristics for locomotor performance. Further and most significantly, the group gave rise independently to two major phylogenetic radiations of endotherms, the mammals and the birds (Fig. 1). These animals are very different in their metabolic and exercise capacities than were their reptilian progenitors, and the evolution of their endothermic condition has been a topic of sustained interest and speculation (reviewed in Bennett, 1991). In these deliberations, the metabolic and exercise status of extant reptiles have figured prominently. Surviving reptilian groups are diversified phylogenetically, and it has been argued (Bennett and Ruben, 1979) on the principle of parsimony that traits that are widely shared among extant reptilian taxa are representative of the ancestral physiological condition from which mammalian and avian endothermy evolved. For example, low resting metabolic rates are characteristic of all extant amphibian and reptilian taxa. These are in contrast to those of birds and mammals, which are 6- to 10-fold greater, even at the same body size and temperature (Benedict, 1932; Dawson and Bartholomew, 1956; Bennett and Dawson, 1976). This greater metabolic rate, coupled with greater insulation, is the physiological basis for endothermy in these groups. Although it has been proposed (Gardiner, 1982) that birds and mammals shared a common endothermic ancestor, the distribution of relevant physiological traits among extant reptiles makes this proposition exceedingly unlikely (Fig. 1). If this were the case, low levels of metabolism would have to have re-evolved at least three times among extant reptilian lineages. More probable is the independent development of

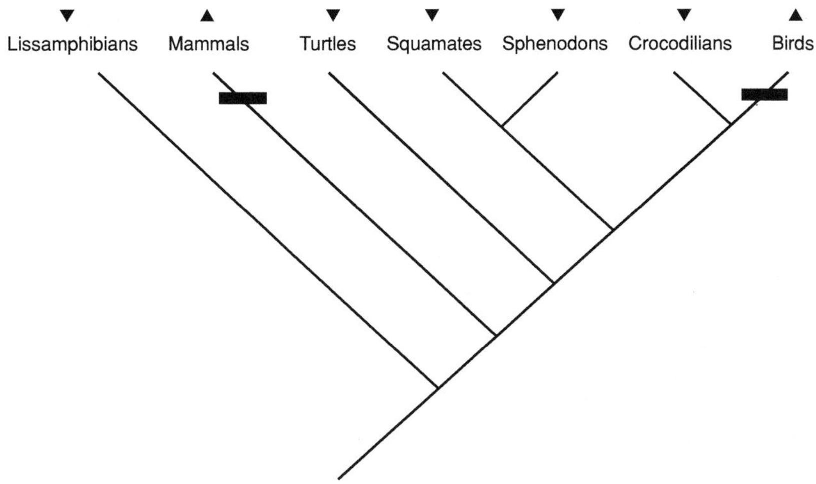

FIG. 1. A classification of extant tetrapod groups (from Gauthier *et al.*, 1988a,b). Low resting metabolic rate is indicated by a downward-pointing triangle; a high metabolic rate by an upward-pointing triangle. Solid bars indicate the probable independent acquisition of high metabolic rate and endothermy. If high metabolic rate was a shared primitive character of mammals and birds, low metabolic rate would have had to have repeatedly re-evolved in the extant reptilian groups.

endothermy in mammals and birds. Such discussions of the macroevolution of physiological characters of mammals and birds, as well as dinosaurs (Bakker, 1971, 1986) and mammallike reptiles (Bennett and Ruben, 1986), have consistently taken those of extant reptiles into account.

Phylogenetic history is increasingly recognized as an important influence on the structure and function of extant organisms (e.g., Lauder, 1981; Felsenstein, 1985; Huey, 1987; Harvey and Pagel, 1991). Comparative studies have begun to incorporate phylogenetic information into the analysis of morphological and physiological characters, attempting to determine ancestral and derived condition, directionality of evolution, and adaptive nature of the character. Analyses of reptilian activity (e.g., Huey and Bennett, 1987; Losos, 1990a,b) were among the first functional studies to incorporate and take advantage of this historical perspective. The availability of excellent phylogenies for many reptilian groups greatly facilitates this approach and ensures the continued use of reptiles in future studies of this type.

A further feature of reptiles that has prompted interest in their activity capacities is the diversity of their locomotor structures. Crocodilians and many lizards have a generalized tetrapodal limb structure that has served as a model for that primitive condition. In addition, the diversity within the group has provided an opportunity to study the consequences of specialization for locomotor performance and energetics. Limb reduction or elimination, for instance, has occurred several times in the group. Snakes in particular have been proved popular study animals for examining the locomotor consequences of limblessness (e.g., Jayne, 1986). Other specialized features influencing locomotor ability include the boxlike shell of turtles, the zygomorphic limb structure of chameleons, and toe-pad specializations of geckos. These all invite investigation of their performance and energetic consequences for locomotion.

In addition to the diversity of their locomotor morphology, reptiles are also ecologically diverse in their habits and habitats, exhibiting a variety of different types and levels of activity under natural circumstances. Some species remain immobile for long periods and have punctuated episodes of very intense activity ("sit-and-wait foragers"), and some are continuously active for long periods ("wide foragers"). Some are active at low and some at high temperatures. Some are terrestrial, some are arboreal, some are aquatic, and some are naturally active in more than one of these habitats. In addition, many reptiles are diurnally active and highly visible and occur locally in great numbers, facilitating observations on their natural behavior and population biology. The wealth of information on reptilian ecology and their popularity in ecological studies make it much easier to place laboratory-based observations in an appropriate natural context.

There are thus many reasons that biologists have used reptiles in studies of locomotor performance and activity, reflected in the literature in a wide variety of observations, motivations, and interpretive contexts. This will not be an extensive and detailed review of all such studies. Rather it will attempt to examine the breadth of topics that physiologists have examined regarding active reptiles and the major contributions of these studies to the field of exercise physiology. Previous reviews are available on reptilian energetics (Bennett and Dawson, 1976; Waldschmidt *et al.*, 1987) and activity physiology (Bennett, 1982, 1983; Gatten, 1985; Gleeson and Bennett, 1985; Hertz *et al.*, 1988; Gleeson, 1991).

II. Endurance Activity

The first comprehensive comparative studies on active animals were done on fish by Fry, Brett, and their co-workers (e.g., Fry, 1947; Brett,

1964). This research involved the measurement of oxygen consumption over a range of sustainable swimming speeds, as well as determination of the maximum sustainable speed ("critical swimming speed"). The influence of such factors as body size and temperature on cost and limits of performance figured prominently in this research. Comparative mammalian studies that followed (e.g., Taylor *et al.*, 1970) continued to emphasize energy expenditure as a function of speed, with concomitant analysis of the influence of body size. Early work on the physiology of active reptiles was much influenced by these studies, and reptilian research therefore initially followed similar themes.

The first studies on reptilian activity involved the measurement of aerobic scope, the difference between standard and maximal oxygen consumption. Fry (1947) proposed this as an index of the sustainable work capacity of an organism, and subsequent comparative studies have generally supported this association. Reptilian studies concentrated on species differences in aerobic scope and found a positive correlation between activity level of the species and aerobic scope (e.g., Tucker, 1967; Ruben, 1976; John-Alder *et al.*, 1986) (e.g., Fig. 2). Maximal aerobic scope generally, although not invariably, was found to correlate with preferred body temperatures, so that work capacity is generally maximized under conditions experienced during field ac-

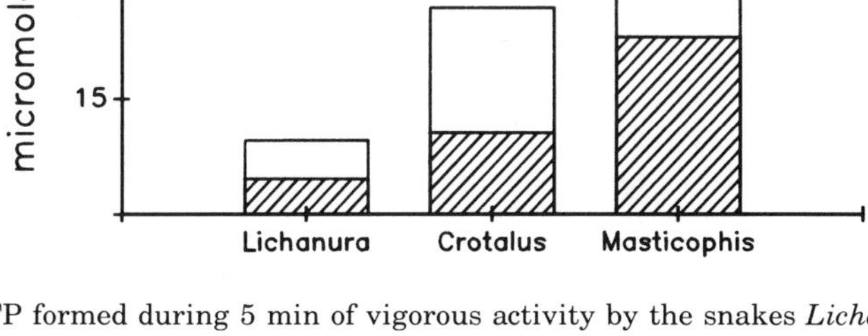

FIG. 2. ATP formed during 5 min of vigorous activity by the snakes *Lichanura roseofusca* (rosy boa), *Crotalus viridis* (western rattlesnake), and *Masticophis flagellum* (coachwhip) (body temperatures = 32–35°C). The boa is very slow and sluggish, the rattlesnake undertakes intense bursts of striking activity, and the coachwhip is an actively foraging predator. The aerobic contribution is measured by oxygen consumption (hatched bar) and the anaerobic (open bar) by lactic acid formation. Data from Ruben (1976).

tivity. Studies of scope, which frequently elicited maximal oxygen consumption through handling or electric shock, gave way to more detailed examinations of locomotor energetics involving treadmill exercise (e.g., Moberly, 1968b; Bakker, 1972). This latter work has been a crucial contribution to our understanding of the energetics of locomotion of animals in general.

A. Cost of Locomotor Activity

Energy expenditure of animals walking or running over a range of sustainable speeds can be adequately described with a linear regression (Fig. 3A). The slope of this function is termed the net cost of transport (Taylor *et al.*, 1970; Schmidt-Nielsen, 1972) and is expressed as energy utilized to move a unit body mass a unit distance. Bakker (1972) first demonstrated and subsequent studies (e.g., John-Alder *et al.*, 1986) have confirmed that this net cost of transport is indistinguishable in mammals and reptiles. It had previously been assumed that the sprawling-limbed posture of ectothermic vertebrates was energetically inefficient, a condition that would have been reflected in a greater net cost of transport. Bakker's data made it clear that this aspect of locomotor energetics is independent of posture. Net cost of transport was not modified during, and thus could not have been a selective factor in, the evolution of endothermy in mammals. Later comparative studies involving other vertebrates and invertebrates (summarized in Full *et al.*, 1990) found that the net cost of transport for terrestrial locomotion may be generally described by a single allometric relation (body mass exponent, b, = -0.3), regardless of body plan, limb number, or phylogenetic affinity. [It should be born in mind, however, that the description of terrestrial locomotor costs by a common allometric equation is not a functional explanation of those costs. Biomechanical and physiological factors that underlie this relationship (e.g., Kram and Taylor, 1990) still need to be elucidated.]

Several studies concerning locomotor costs of reptiles, in addition to those cited here, have been influential in our more general understanding of locomotor energetics. Of particular interest in this regard has been the energetic effects of the evolutionary reduction or elimination of limbs. An early report measuring the cost of crawling in snakes (Chowdrow and Taylor, 1973) suggested a lower than anticipated net cost of transport. Because limb reduction has developed repeatedly in different groups of tetrapods, these data suggested that one explanation for its evolution might have been reduction of locomotor costs and consequent energetic economy. A subsequent study, however, found net

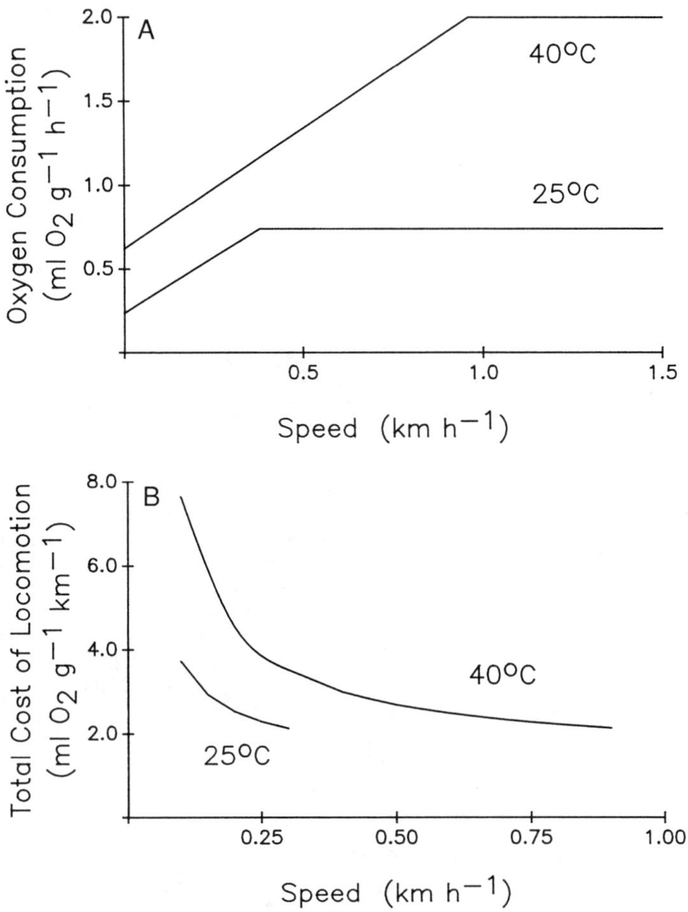

FIG. 3. Oxygen consumption by the lizard *Dipsosaurus dorsalis* walking at different speeds with body temperatures of 25° and 40°C. (A) Oxygen consumption increases up to maximal aerobic speed (0.38 km h^{-1} at 25°C; 0.96 km h^{-1} at 40°C). Net cost of transport (slope of the line) is temperature independent and equals 1.4 ml O_2 g^{-1} h^{-1}. (B) Total cost of locomotion (energy used to walk 1 km) decreases with increasing speed. It is greater at 40° than at 25°C at any common sustainable speed, and is temperature independent (= 2.1 ml O_2 g^{-1} km^{-1}) at maximal aerobic speed. Data from John-Alder and Bennett (1981).

costs of lateral undulation and concertina locomotion to be equal to or greater than those of limbed locomotion (Walton *et al.*, 1990), and hence no particular energetic advantage apparently accrued during the evolution of limblessness (Fig. 4). Sidewinding, a relatively uncommon locomotor mode even among snakes, apparently does have a lower

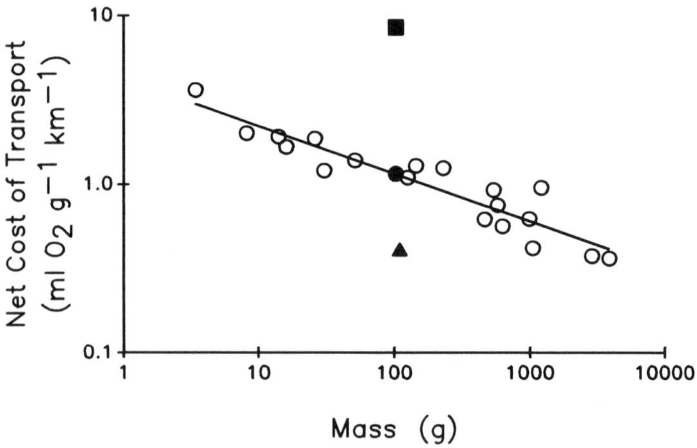

FIG. 4. Net cost of transport (slope of oxygen consumption as a function of speed) of lizards and snakes of different size. Open circles = 15 different lizard species (summarized in John-Alder *et al.*, 1986); solid circle and square = lateral undulation and concertina locomotion (respectively) by the snake *Coluber constrictor* (Walton *et al.*, 1990); solid triangle = sidewinding by *Crotalus cerastes* (Secor *et al.*, 1992). Regression line is for lizard species only (John-Alder *et al.*, 1986).

than anticipated net cost of transport, but substantially higher total costs at low speeds (Secor *et al.*, 1992).

Another topic of interest to comparative exercise physiologists is the profound difference in transport cost evident in locomotion in different media. Net costs of transport of flying animals are only about 30% and those of swimmers are only about 10% those of terrestrial walkers or runners (Schmidt-Nielsen, 1972). These comparisons, however, are made among animals of extremely different body form and phylogenetic history, such as birds, fish, and mammals, and it is difficult to disengage other morphological and phylogenetic effects on this relationship from those associated with locomotor costs alone. Marine iguanas are lizards that naturally engage in both walking and swimming behavior, so it was possible (Gleeson, 1979, 1980; Vleck *et al.*, 1981) to measure differential locomotor costs within a single species and thus reduce extraneous effects arising from intergroup comparisons (Fig. 5). Terrestrial locomotion for these lizards proved to be as expensive as in other terrestrial walkers. Swimming is in fact substantially less costly than walking for these animals, although it is not as inexpensive as it is in fish, presumably because of the energy involved in surface wake formation. Differences in the cost of transport among media are not, therefore, simply due to differences in body form in

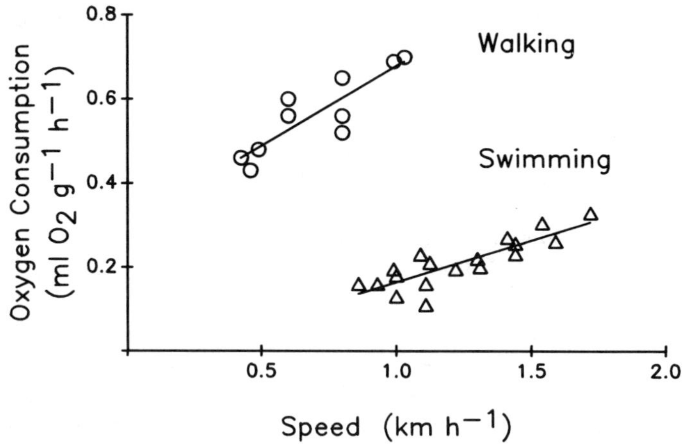

FIG. 5. Oxygen consumption during walking (body temperature = 35°C) and swimming (body temperature = 25°C) in adult marine iguanas. Locomotor costs are reduced (net cost of transport = 0.185 versus 0.373 ml O_2 g^{-1} km^{-1}) and endurance is increased by swimming in comparison to walking. Data from Gleeson (1979).

different taxa or factors such as differential amount of locomotor musculature. Rather they appear to involve more fundamental factors inherent in the media themselves, such as differential buoyancy and resistance. These studies demonstrate the past utility of reptiles in studies of locomotor energetics, and we may anticipate that reptiles will continue to play an important role in the future development and testing of bioenergetic models.

Locomotor costs are not completely accounted for by measurement of net cost of transport alone. Total metabolic rates also include those costs estimated by the extrapolation of energy use to zero speed and are variously assumed to include maintenance and postural factors (Fig. 3A) (Taylor *et al.*, 1970). These may make a substantial difference in comparisons of locomotor energetics. For instance, although net costs of transport are equal in reptiles and mammals, total metabolic rates of similar-sized animals walking at similar speeds are approximately 6- to 10-fold greater in mammals (Bennett and Ruben, 1979). Likewise, although net costs of transport are temperature independent in reptiles, maintenance costs are not, and it is more expensive for an animal to walk at a given speed with a higher body temperature (Fig. 3B). Total locomotor costs are the ones that actually must be met from the energy budget of an animal, and hence are the factors to be considered is estimating locomotor costs in natural environments.

B. Endurance and Maximal Oxygen Consumption

Walking or running is not indefinitely sustainable at all speeds at which an animal is capable of moving. As speed increases, a limit is reached, above which exhaustion ensues rapidly. This limit is termed the maximal aerobic speed and coincides with the attainment of maximal oxygen consumption during exercise (Fig. 3A) (Bennett and John-Alder, 1984). Maximal aerobic speed is correlated with maximal oxygen consumption across body temperatures within a species and these factors are positively associated among species (Bennett, 1982; Bennett and John-Alder, 1984). Speeds above maximal aerobic require sources of energy in addition to those provided by aerobic metabolism, involving in reptiles the catabolism of high-energy phosphate compounds and/or the glycolytic formation of lactic acid (Bennett, 1978, 1989). Both resting and maximal levels of oxygen consumption of reptiles are low in comparison to the costs of terrestrial locomotion, and hence maximal aerobic speeds are relatively modest, typically approximating only 1 km h^{-1} (Garland, 1982). Maximal speeds during rapid bursts of activity may exceed this value by 10- to 20-fold (Bennett, 1980, 1982), and hence only a fraction of the potential locomotor repertoire of reptiles can be accessed and sustained aerobically. Low body temperatures exacerbate this condition by decreasing maximal levels of oxygen consumption and hence decreasing maximal aerobic speeds (Bennett and John-Alder, 1984). Reptiles consequently have a much lower capacity for sustainable activity than do birds or mammals, and this has been hypothesized as one of the factors associated with the evolutionary development of endothermy in these groups (Regal, 1978; Bennett and Ruben, 1979). Reptiles are confined by their physiological capacities to undertake rather slow activities that can be met with their modest aerobic powers or to engage in bursts of rapid movement that quickly result in exhaustion.

If maximal oxygen consumption constrains sustainable activity levels, what limits maximal oxygen consumption? There have been no systematic examinations of this limitation in reptiles, in contrast to extensive investigations undertaken in mammals (Taylor and Weibel, 1981; Weibel *et al.*, 1991). This latter work has been done in the investigation of the concept of symmorphosis, the hypothesis that no single morphological or physiological system limits a complex function, such as maximal oxygen consumption. Rather design limits are hypothesized to be met in all components of the system simultaneously. This hypothesis and interpretation of data supporting it have been controversial (Lindstedt and Jones, 1987; Dudley and Gans, 1991), and it would seem well worthwhile to undertake a similar investigation in another group of animals to test the hypothesis further. Reptiles would be a very logical choice for such an examination, because of the relative ease of investigating their physiology during sustained locomotion.

Studies on maximal oxygen consumption in reptiles have focused on the role of particular systems in providing oxygen delivery during activity. It has been proposed (Carrier, 1987, 1989, 1990) that ventilation of the lungs in lizards is incompatible with locomotion because of competing functional demands on the axial musculature during locomotion and ventilation. Hence ventilation must be restricted while animals are walking or running, forming a limitation on oxygen uptake during activity, particularly at high speeds. Other measurements on active lizards (Gleeson *et al.*, 1980; Mitchell *et al.*, 1981a,b), however, have found that breathing is sustained during lower-speed locomotion, even at velocities greater than maximal aerobic speed, at which maximal oxygen consumption is attained. Both alveolar and arterial oxygen partial pressures remain high at speeds above maximal aerobic (Fig. 6), indicating that ventilation is quite adequate for oxygen supply over this entire speed range. These observations indicate that ventilation is not necessarily compromised during locomotion and is not a limiting factor in oxygen transport at or above maximal aerobic speed in these animals. If lizards do not ventilate while running at high speeds, it may be because other portions of the oxygen transport systems are already operating at maximal capacity (Fig. 7) and continued ventilation would not increase oxygen uptake. Human sprinters, for instance, frequently avoid breathing during their runs.

In contrast to ventilation, cardiac function may be a factor limiting maximal oxygen in consumption. Limits of cardiovascular transport capacities appear to be attained at maximal aerobic speed, that is, maximal heart rate, stroke volume, and oxygen extraction are reached (Fig. 7) (Gleeson *et al.*, 1980). Several interindividual studies of reptilian populations have shown significant relationships between cardiovascular factors (e.g., ventricular mass, blood hemoglobin concentration) and maximal oxygen consumption (Schall *et al.*, 1982; Garland, 1984; Garland and Else, 1987; Garland and Bennett, 1990). The latter study also found a significant genetic, as well as phenotypic, correlation between ventricular mass and maximal oxygen consumption in garter snakes. Potentially, therefore, cardiovascular function may be a limiting factor in oxygen transport and endurance capacity in reptiles. The physiological limitations on maximal oxygen consumption in this group are certain to remain a topic of interest and scrutiny, as more comprehensive investigations and techniques are directed to the problem.

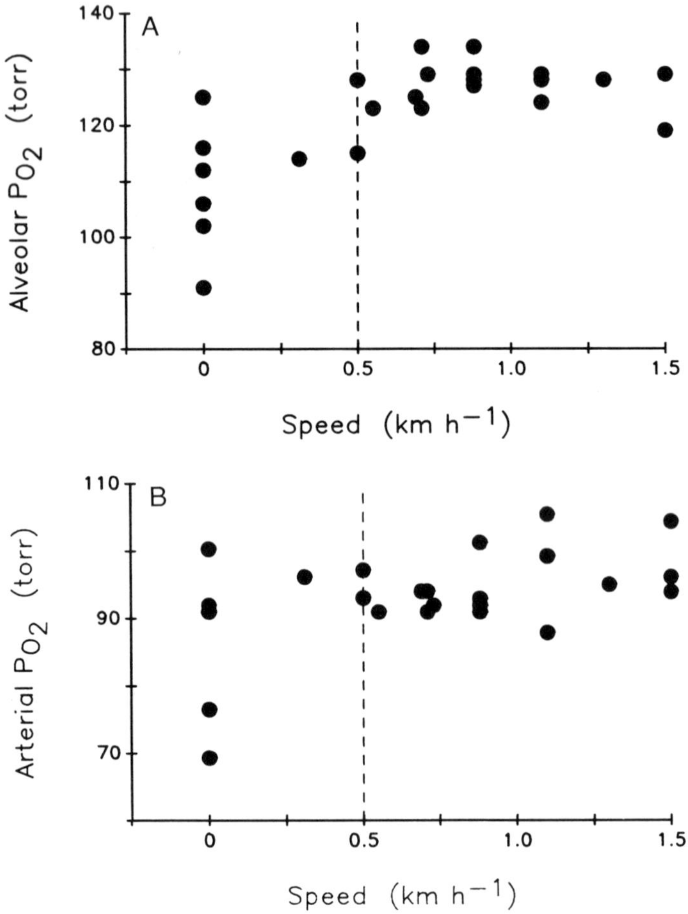

FIG. 6. Alveolar (a) and arterial (b) oxygen partial pressure in the lizard *Iguana iguana* walking at different speeds (body temperature = 35°C). Maximal aerobic speed (vertical dashed line) is 0.5 km h^{-1}. Alveolar and arterial oxygen partial pressures remain high even at 1.5 km h^{-1}, indicating that ventilation is adequate to oxygenate the blood. Ventilation therefore does not limit oxygen uptake. Recalculated from data of Mitchell *et al.* (1981a,b).

C. Ecological and Evolutionary Implications of Endurance

Do reptiles commonly utilize high levels of aerobically supported activity under natural conditions? Most reptiles probably do not engage daily in sustained levels of endurance activity (Hertz *et al.*, 1988). However, in some specific circumstances, such as egg laying in sea

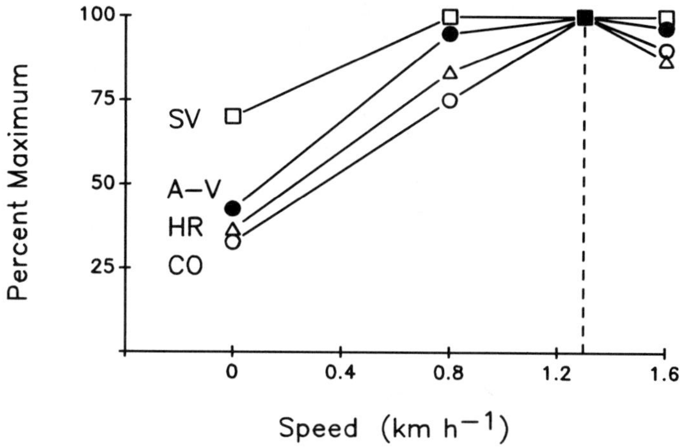

FIG. 7. Cardiovascular function in the lizard *Varanus exanthematicus* walking at different speeds (body temperature = 35°C). Maximal aerobic speed (vertical dashed line) is 1.3 km h^{-1}. Values are plotted as a percentage of maximal value measured at any speed: maximal heart rate (open triangles) = 115 beat min^{-1}, maximal stroke volume (open squares) = 3.0 ml $beat^{-1}$, maximal cardiac output (open circles) = 350 ml min^{-1}, and maximal oxygen extraction (A–V difference) (solid circles) = 0.61 ml O_2 ml $blood^{-1}$. Cardiovascular function is maximized at maximal aerobic speed. Data recalculated from Gleeson *et al.* (1980).

turtles (Jackson and Prange, 1979) or physiological thermoregulation during incubation in pythons (Hutchison *et al.*, 1966; Vinegar *et al.*, 1970), sustained levels of oxygen consumption are quite high, probably at or approaching maximal oxygen consumption. Interspecific comparisons generally show a positive association between aerobic capacity and level of foraging intensity, but their interpretation is complicated by phylogenetic diversity (Hertz *et al.*, 1988). For instance, varanid and teiid lizards generally have greater maximal oxygen consumption and endurance than do iguanid and agamid lizards (e.g., Bartholomew and Tucker, 1964; Gleeson, 1981; Anderson and Karasov, 1981; John-Alder *et al.*, 1983; Garland, 1993). These former groups tend to have more continuous patterns of foraging than do the latter. Such comparisons, however, are complicated by many factors. Comparative measurements among closely related species that differ in level of natural activity are more useful in eliminating extraneous factors from these comparisons. One such study compares the ecology, behavior, and physiology of two closely related (congeneric) lacertid lizards (Table I) (Nagy *et al.*, 1984; Huey *et al.*, 1984, Bennett *et al.*, 1984). *Eremias lugubris* is a continuous forager and *E. lineoocellata* is an ambush

TABLE I

FIELD METABOLIC RATES AND BEHAVIORAL AND PHYSIOLOGICAL CAPACITIES OF TWO SMALL LACERTID LIZARDS[a]

	Eremias lugubris	*Eremias lineoocellata*
Foraging mode	Widely foraging	Sit-and-wait
Field metabolic rates		
Cost of foraging (J h^{-1})	214	44
Daily energy use (J day^{-1})	800	544
Behavioral capacities		
Sprint speed (m s^{-1})	1.58	2.63
Maximal exertion (m)	70.8	29.9
Endurance (min at 0.5 km h^{-1})	>30	7
Physiological capacities		
Maximal oxygen consumption (ml O_2 g^{-1} h^{-1})	3.22	2.49
Anaerobic scope (mg lactate g^{-1} min^{-1})	2.12	2.56
Anaerobic capacity (mg lactate g^{-1})	1.40	1.81
Heart mass (mg)	7.7	5.2
Hematocrit (%)	30.1	24.4

[a] Body temperature = 35–37°C. All differences between species are significant. Data from Bennett *et al.* (1984), Huey *et al.* (1984), and Nagy *et al.* (1984).

predator, and both live together in the same desert habitat. The former has higher maximal oxygen consumption, endurance, and field metabolic rate than does the latter. Field metabolic rate measurements indicate that *E. lugubris* forages at or close to its maximal aerobic limits. Its greater maximal oxygen consumption is correlated with cardiovascular factors, but not with differences in skeletal muscle physiology. Greater aerobic capacities as detected in the laboratory for this species may, therefore, both reflect and permit its greater level of field foraging intensity. Measurements of daily field metabolic rate indicate that lizards utilize about 100–200% more energy that can be accounted for by their resting (or standard) metabolic rates alone (Bennett and Nagy, 1977; Nagy, 1982).

How important are these capacities for sustainable activity to reptiles in nature? Are high capacities for endurance useful in promoting survival or enhanced growth and reproduction? Research on these questions has only recently begun and is confined to few studies. It has concentrated on the correlation between endurance and survivorship among individual animals in natural populations (garter snakes, Jayne and Bennett, 1990b; sceloporine lizards, unpublished observations of R. B. Huey and others cited in Bennett and Huey, 1990). None of these studies has found a significant association between endurance and survivorship (although the correlation is marginally significant, $0.1 > P > 0.05$, for older snakes), indicating no detectable natural selection on endurance. None of the species examined are widely foraging or highly active reptiles, and the generality of these results is unknown. Hertz *et al.* (1988) consider several alternative explanations for the maintenance of physiological capacities that may not be used frequently under field conditions, such as endurance. Among those proposed are the relatively rare (and therefore generally unobserved) nature of an appropriate selective event or the possibility that the trait in question is correlated with performance that is only adequate in some other condition (e.g., at lower body temperatures). However, the survivorship studies cited should, within statistical limits, encompass these possibilities in natural populations (Bennett and Huey, 1990). Maybe circumstances requiring the trait are truly rare, that is, not encountered every generation but only once in every ten or hundred generations, and therefore not detected easily even in multiyear studies. At present, the significance and maintenance of endurance capacities in reptiles are not understood.

III. Burst and Exertion Activity

A. Anaerobic Glycolysis and Metabolic Support

Most reptiles are capable of very intense, short-lived bursts of activity. These bouts are not supported through oxygen consumption, but rather by anaerobic metabolic pathways. As in aerobic exercise physiology, comparative studies of metabolism during burst activity were first undertaken in fish, particularly by Black and his co-workers (1961). This research showed that intense activity is accompanied by the catabolism of muscle glycogen and the appearance of large quantities of lactic acid in the blood. Studies on reptiles (e.g., Austin *et al.*, 1927; Moberly, 1968a; Gatten, 1975) also found very large increments in blood lactate as a result of brief struggling activity. [It is now clear that use of anaerobic glycolysis in support of intense struggling is nearly universal among the vertebrates and is probably a primitive character of the chordate and perhaps deuterostome lineage (Ruben and Bennett, 1980; Bennett, 1991).] However, measurement of changes in blood lactate concentration in exercise physiology has interpretive limitations. Because of the compartmentalization of lactate within the body and long lags associated with intercompartmental transfers (e.g., muscle to blood), blood lactate concentrations cannot be used to quantify the total amount of anaerobic metabolism or its contribution to activity energetics. The analytical development of whole-body lactate analysis (Bennett and Licht, 1972) permitted the quantification of the anaerobic energetic contribution to activity. It became possible to estimate directly the contribution of anaerobiosis to activity and to measure the maximal rate of lactic acid formation (anaerobic scope) and the amount formed during activity to exhaustion (anaerobic capacity). It also became possible to compare simultaneously aerobic and anaerobic metabolic support for any behavior or sequence of behaviors of brief duration. In the first experiments undertaken, the level or amount of burst activity was not quantified: reptiles were stimulated manually or electrically. Later, locomotor performance was measured directly as maximal speed attained on a racetrack ("burst speed," Bennett, 1980; Hertz *et al.*, 1982) or as distance covered during several minutes or to exhaustion ("maximal exertion" *sensu* Bennett, 1989) (Bennett, 1980).

Bursts of intense activity in reptiles are fueled primarily through anaerobic glycolysis, resulting in lactic acid formation. For example, during burst activity in a small lizard, about 60% of energy utilized is derived from lactate formation, about 6% from catabolism of adenosine triphosphate and creatine phosphate, and only 1% from other anaerobically formed compounds; the balance is aerobic (Table I). High levels of lactate formation during activity have been found in all groups of extant reptiles examined (squamates, turtles, and crocodilians) (reviewed in Bennett, 1978, 1982; Gatten, 1985), with blood lactate concentrations sometimes exceeding 50 m*M* (Bennett *et al.*, 1985). Simultaneous measurement of oxygen consumption and lactic acid formation (Table II and Fig. 8) indicates that the latter accounts for 60–90% of the ATP formed during 2-min bouts of intense activity, aerobic metabolism playing a relatively minor energetic role (Bennett and Dawson, 1972; Bennett and Gleeson, 1976). This balance results from the combination of high anaerobic and low aerobic capacities in reptiles. In mammals, in contrast, 75–92% of the ATP equivalents formed during 2 min of burst activity are derived aerobically (Ruben and Battalia, 1979). Anaerobic metabolic support in reptiles assumes an even greater fraction of total energy production as body temperature decreases (Fig. 8). The aerobic contribution is more strongly temperature dependent, with a temperature coefficient (Q_{10}) of approximately 2 (reviewed in Bennett, 1982), in comparison to temperature coefficients of 1.1–1.3 for lactic acid formation (Bennett and Licht,

TABLE II

ATP EQUIVALENTS GENERATED BY THE LIZARD *SCELOPORUS OCCIDENTALIS* DURING 2 MIN OF BURST ACTIVITY[a]

	Micromole ATP (g body mass^{-1})	Percent (total)
Aerobic		
Oxygen consumption	12.7	33
Anaerobic		
Production of		
Lactic acid	22.5	59
Pyruvic acid	0.2	<1
Alanine	0.2	<1
Succinic acid	0	0
Catabolism of		
ATP	0.9	2
Creatine phosphate	1.6	4

[a] Body temperature = 35°C. Unpublished data from A. F. Bennett and T. T. Gleeson.

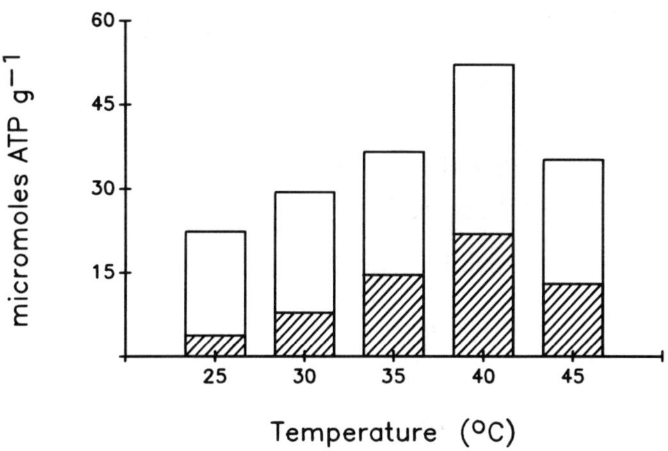

FIG. 8. Thermal dependence of aerobic and anaerobic contributions to energy production during 2 min of burst activity by the lizard *Dipsosaurus dorsalis*. Aerobic metabolism (hatched bar) is measured by oxygen consumption; anaerobic contribution (open bar) by whole-body lactic acid accumulation. The aerobic contribution has a stronger thermal dependence (Q_{10} = 3.23 versus anaerobic Q_{10} = 1.39 from 25° to 40°C), resulting in a diminished percentage contribution at low body temperatures. Data from Bennett and Dawson (1972).

1972; Bennett, 1982). Anaerobiosis thus ensures a capacity for intense exercise and resulting escape or pursuit behavior over a relatively wide range of body temperatures, even during seasons of low activity (Gatten *et al.*, 1988). For this reason and for its very rapid mobilization and intensity, anaerobic glycolysis has been referred to as the Smith and Wesson, the great equalizer of the ectotherms and endotherms (Coulson, 1979).

The large amounts of lactic acid formed during activity in reptiles are highly physiologically disruptive (see Bennett, 1982, for a discussion). Blood and tissue can experience pronounced acidosis, with blood pH values declining as low as 6.79 in snakes (Ruben, 1979) and 6.42 in crocodiles (Bennett *et al.*, 1985). This acidosis may disrupt enzyme (Hochachka and Somero, 1984) and hemoglobin (Bennett, 1973) function and perturb blood calcium homeostasis (Ruben and Bennett, 1981). The relatively low ATP yield per carbohydrate molecule catabolized requires extensive depletion of muscle glycogen (Gleeson, 1991). These physiological disruptions are associated with pronounced and prolonged exhaustion after a bout of intense exertion. Physiological factors responsible for this exhaustion are not known. Physiological recovery by reptiles from intense activity (reviewed by Gleeson, 1991) takes a long time, as much as several hours, and rate of recovery is greatly retarded by low body temperatures.

B. Anaerobic Metabolism under Natural Conditions

In view of its potential for physiological disruption and exhaustion, we might well consider the extent to which anaerobic metabolism is normally used during natural activities by reptiles (e.g., Hertz *et al.*, 1988). A number of studies have examined the extent of lactic acid formation during staged behaviors in the laboratory or under natural conditions in the field. In several species of lizards sampled in the field without regard to their prior behavior, lactate concentrations are moderately elevated in comparison to resting or preemergence values (Bennett *et al.*, 1981; Pough and Andrews, 1985a), indicating general increase associated with routine activity. Some species show no such elevation (Pough and Andrews, 1985a). Lactic acid is formed during conspecific agonistic encounters by lizards (Bennett *et al.*, 1981; Pough and Andrews, 1985a; Wilson and Gatten, 1989; Wilson *et al.*, 1990), and accumulation is correlated with the intensity of the aggression (e.g., Fig. 9). Moderate amounts of glycolysis are observed in snakes feeding on salamanders (Feder and Arnold, 1982); significant but relatively lower levels have been measured in lizards eating insects (Pough

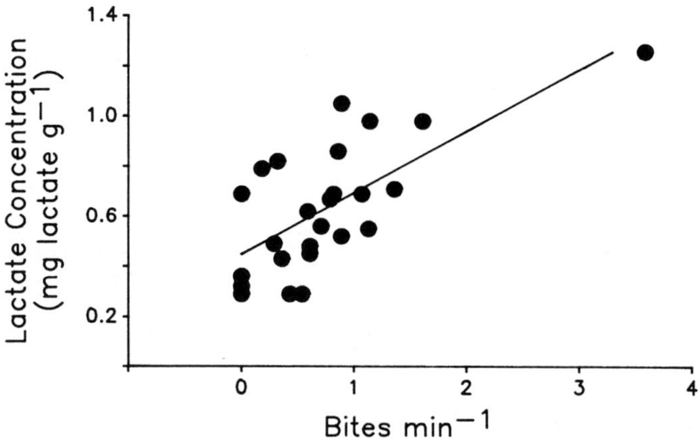

FIG. 9. Lactate contents of resident lizards (*Sceloporus jarrovi*) resulting from territorial defense against an experimentally placed intruder. Intensity of defense activity was measured as the rate of bites inflicted on the intruder. Intensity is directly correlated with lactate formation. Lizards at morning emergence and during routine daily activity have average lactate concentrations of 0.26 and 0.31 mg lactate g^{-1}, respectively. Data redrawn from Pough and Andrews (1985a).

and Andrews, 1985a,b). Lactate accumulation occurs during egg laying by sea turtles in the field (Jackson and Prange, 1979) and in hatchling sea turtles during digging out of the nest, crawling across the beach, and swimming through the surf (Dial, 1987; Baldwin *et al.*, 1989). Turtles hibernating underwater utilize anaerobic metabolism (Gatten, 1981, 1985; Ultsch and Jackson, 1982), accumulating phenomenal concentrations of lactate (up to 5.5 mg g^{-1}; Gatten, 1981). Lactate accumulation also occurs in reptiles during forced diving in the laboratory, but is usually avoided during free diving behavior in the field (reviewed by Seymour, 1982; Gatten, 1985), a situation similar to that observed in diving mammals and birds. Hertz *et al.* (1988) concluded that reptiles frequently utilize anaerobic metabolism during natural activities, but rarely do so to the point of exhaustion. Anaerobic metabolism appears to be particularly useful in and used for fueling intense activities of short duration but great ecological importance (e.g., territorial defense, nest emergence).

C. Ecological and Evolutionary Consequences

What are the ecological and evolutionary consequences of burst speed and maximal exertion capacities for reptiles? Both factors have been demonstrated to be heritable (van Berkum and Tsuji, 1987; Garland, 1988; Tsuji *et al.*, 1989; Jayne and Bennett, 1990a,b), indicating a genetic basis and potential to respond to natural selection (Bennett and Huey, 1990). In staged laboratory encounters, individual male lizards with greater burst speeds (but not endurance) were dominant in social interactions (Garland *et al.*, 1990). Reptiles fleeing predators in nature appear to do so at the maximal speeds they can attain (Greene, 1988; Hertz *et al.*, 1988). For example, marine iguanas observed while naturally escaping from hawks ran as fast as when they were pursued experimentally (Gleeson, 1979, 1980). Burst speed capacity (and body-size-adjusted burst speed capacity) has been significantly and positively associated with survivorship among individuals in natural populations of garter snakes (Jayne and Bennett, 1990b) and tree lizards (Miles, 1989), indicating that it is under directional selection in these populations. Other studies (R. B. Huey and co-workers, as cited in Bennett and Huey, 1990) on fence lizards and canyon lizards have found no significant association between survivorship and speed. Thus, the ecological and microevolutionary significance of these factors apparently may vary among populations, at least among species, and perhaps with time and environment within a single population.

IV. Summary

From the vantage point of thirty years of study, we can sketch the general features of activity capacity and performance ability in reptiles. Extant reptilian groups all share low levels of maintenance metabolism and ectothermy, with their consequent advantages (Pough, 1980) and disadvantages. Among the latter is a limited capacity to expand aerobic metabolism, limited in comparison to the relatively great costs of terrestrial locomotion. Particularly at low body temperatures, reptiles outstrip their aerobic capacities with any exercise more intense than a slow walk. Anaerobic metabolism, particularly anaerobic glycolysis, can be used to fuel bursts of intense activity. As a consequence, however, physiological disruption and exhaustion are entailed. Under field conditions, many reptiles alternate long periods of quiescence or slow movement with very brief bursts of exertion. Other ectotherms with a similar pattern of metabolism have been shown thereby to extend performance beyond that supportable by either aerobic or anaerobic metabolism alone (Weinstein and Full, 1992). Even with careful alternation between these metabolic modes, reptiles remain particularly prone to exhaustion during vigorous activity, as least as judged by our mammalian frame of reference. Their capacities for burst activity and exertion have been shown, at least in some species, to be important determinants of their natural survival.

Acknowledgments

I thank Todd Gleeson and Gordon Mitchell for help with data preparation and Alistair Cullum and Craig Frank for assistance with figure preparation. Financial support was provided by National Science Foundation Grants BSR-8918054 and IBN-9118346.

References

Anderson, R. A., and Karasov, W. H. (1981). Contrasts in energy intake and expenditure in sit-and-wait and widely foraging lizards. *Oecologia* **59,** 67–72.

Austin, J. H., Sunderman, F. W., and Camack, J. G. (1927). Studies in serum electrolytes. II. The electrolyte composition and the pH of serum of a poikilothermous animal at different temperatures. *J. Biol. Chem.* **72,** 677–685.

Bakker, R. T. (1971). Dinosaur physiology and the origin of mammals. *Evolution (Lawrence, Kans.)* **25,** 636–658.

Bakker, R. T. (1972). Locomotor energetics of lizards and mammals compared. *Physiologist* **15,** 76.

Bakker, R. T. (1986). "The Dinosaur Heresies." Wm. Morrow, New York.

Baldwin, J., Gyuris, E., Mortimer, K., and Patak, A. (1989). Anaerobic metabolism during dispersal of green and loggerhead turtle hatchlings. *Comp. Biochem. Physiol. A* **94A,** 663–665.

Bartholomew, G. A., and Tucker, V. A. (1963). Control of changes in body temperature, metabolism and circulation by the agamid lizard, *Amphibolurus barbatus. Physiol. Zool.* **36,** 199–218.

Bartholomew, G. A., and Tucker, V. A. (1964). Size, body temperature, thermal conductance, oxygen consumption and heart rate in Australian varanid lizards. *Physiol. Zool.* **37,** 341–354.

Benedict, F. G. (1932). The physiology of large reptiles. *Carnegie Inst. Washington Publ.* **425.**

Bennett, A. F. (1973). Blood physiology and oxygen transport during activity in two lizards, *Varanus gouldii* and *Sauromalus hispidus. Comp. Biochem. Physiol. A* **46A,** 673–690.

Bennett, A. F. (1978). Activity metabolism of the lower vertebrates. *Annu. Rev. Physiol.* **40,** 447–469.

Bennett, A. F. (1980). The thermal dependence of lizard behaviour. *Anim. Behav.* **28,** 752–762.

Bennett, A. F. (1982). Energetics of activity in reptiles. *In* "Biology of the Reptilia" (C. Gans and F. H. Pough, eds.), Vol. 13, pp. 155–199. Academic Press, New York.

Bennett, A. F. (1983). Ecological consequences of the activity metabolism of lizards. *In* "Lizard Ecology" (R. B. Huey, E. R. Pianka, and T. W. Schoener, eds.), pp. 11–23. Harvard Univ. Press, Cambridge, MA.

Bennett, A. F. (1989). Integrated studies of locomotor performance. *In* "Complex Organismal Functions: Integration and Evolution in Vertebrates" (D. B. Wake and G. Roth, eds.), pp. 191–202. Wiley, Dahlem Konferenzen, Chichester, England.

Bennett, A. F. (1991). The evolution of activity capacity. *J. Exp. Biol.* **160,** 1–23.

Bennett, A. F., and Dawson, W. R. (1972). Aerobic and anaerobic metabolism during activity in the lizard *Dipsosaurus dorsalis*. *J. Comp. Physiol.* **81,** 289–299.

Bennett, A. F., and Dawson, W. R. (1976). Metabolism. *In* "Biology of the Reptilia" (C. Gans and W. R. Dawson, eds.), Vol. 5, pp. 127–223. Academic Press, New York.

Bennett, A. F., and Gleeson, T. T. (1976). Activity metabolism in the lizard *Sceloporus occidentalis*. *Physiol. Zool.* **49,** 65–76.

Bennett, A. F., and Huey, R. B. (1990). Studying the evolution of physiological performance. *Oxford Surv. Evol. Biol.* **7,** 251–284.

Bennett, A. F., and John-Alder, H. B. (1984). The effect of body temperature on the locomotory energetics of lizards. *J. Comp. Physiol.* **155,** 21–27.

Bennett, A. F., and Licht, P. (1972). Anaerobic metabolism during activity in lizards. *J. Comp. Physiol.* **81,** 277–288.

Bennett, A. F., and Nagy, K. A. (1977). Energy expenditure in free-ranging lizards. *Ecology* **58,** 697–700.

Bennett, A. F., and Ruben, J. A. (1979). Endothermy and activity in vertebrates. *Science* **206,** 649–654.

Bennett, A. F., and Ruben, J. A. (1986). The metabolic and thermoregulatory status of therapsids. *In* "The Ecology and Biology of Mammal-like Reptiles" (N. Hotton, P. D. MacLean, J. J. Roth, and E. C. Roth, eds.), pp. 207–218. Smithsonian Institute, Washington, DC.

Bennett, A. F., Gleeson, T. T., and Gorman, G. C. (1981). Anaerobic metabolism in lizards under natural conditions. *Physiol. Zool.* **54,** 237–241.

Bennett, A. F., Huey, R. B., and John-Alder, H. B. (1984). Physiological correlates of natural activity and locomotor capacity in two species of lacertid lizards. *J. Comp. Physiol.* **154,** 113–118.

Bennett, A. F., Seymour, R. S., Bradford, D. F., and Webb, G. J. W. (1985). Mass-dependence of anaerobic metabolism and acid–base disturbance during activity in the salt-water crocodile, *Crocodylus porosus*. *J. Exp. Biol.* **118,** 161–171.

Black, E. C., Robertson, A. C., and Parker, R. R. (1961). Some aspects of carbohydrate metabolism in fish. *In* "Comparative Physiology of Carbohydrate Metabolism in Heterothermic Animals" (A. W. Martin, ed.), pp. 89–124. Univ. of Washington Press, Seattle.

Brett, J. R. (1964). The respiratory metabolism and swimming performance of young sockeye salmon. *J. Fish. Res. Board Can.* **21,** 1183–1226.

Carrier, D. R. (1987). The evolution of locomotor stamina in tetrapods: Circumventing a mechanical constraint. *Paleobiology* **13,** 326–341.

Carrier, D. R. (1989). Ventilatory action of the hypaxial muscles of the lizard *Iguana iguana*: A function of slow muscle. *J. Exp. Biol.* **143,** 435–457.

Carrier, D. R. (1990). Activity of the hypaxial muscles during walking in the lizard *Iguana iguana*. *J. Exp. Biol.* **152,** 453–470.

Chowdrow, R. E., and Taylor, C. R. (1973). Energetic cost of limbless locomotion in snakes. *Fed. Proc., Fed. Am. Soc. Exp. Biol.* **32,** 422.

Coulson, R. A. (1979). Anaerobic glycolysis: The Smith and Wesson of the heterotherms. *Perspect. Biol. Med.* **22,** 465–479.

Dawson, W. R., and Bartholomew, G. A. (1956). Relation of oxygen consumption to body weight, temperature and temperature acclimation in lizards *Uta stansburiana* and *Sceloporus occidentalis*. *Physiol. Zool.* **29,** 40–51.

Dial, B. E. (1987). Energetics and performance during nest emergence and the hatchling frenzy in loggerhead sea turtles (*Caretta caretta*). *Herpetologica* **43,** 307–315.

Dudley, R., and Gans, C. (1991). A critique of symmorphosis and optimality models in physiology. *Physiol. Zool.* **64,** 627–637.

Feder, M. E., and Arnold, S. J. (1982). Anaerobic metabolism and behavior during predatory encounters between snakes (*Thamnophis elegans*) and salamanders (*Plethodon jordani*). *Oecologia* **53,** 93–97.
Felsenstein, J. (1985). Phylogenies and the comparative method. *Am. Nat.* **125,** 1–15.
Fry, F. E. J. (1947). Effects of the environment on animal activity. *Publ. Ont. Fish. Res. Lab.* **68,** 1–62.
Full, R. J., Zuccarello, D. A., and Tullis, A. (1990). Effect of variation in form on the cost of terrestrial locomotion. *J. Exp. Biol.* **150,** 233–246.
Gardiner, B. (1982). Tetrapod classification. *Zool. J. Linn. Soc.* **74,** 207–232.
Garland, T. (1982). Scaling maximal running speed and maximal aerobic speed to body mass in mammals and lizards. *Physiologist* **25,** 338.
Garland, T. (1984). Physiological correlates of locomotory performance in a lizard: An allometric approach. *Am. J. Physiol.* **247,** R806–R817.
Garland, T. (1988). Genetic basis of activity metabolism. I. Inheritance of speed, stamina and antipredator displays in the garter snake *Thamnophis sirtalis*. *Evolution (Lawrence, Kans.)* **42,** 335–350.
Garland, T. (1993). Locomotory performance and activity metabolism of *Cnemidophorus tigris* in relation to natural behaviors. *In* "Biology of *Cnemidophorus*" (J. W. Wright and L. J. Vitt, eds.), pp. 163–210. Oklahoma Museum of Natural History, Norman, OK.
Garland, T., and Bennett, A. F. (1990). Quantitative genetics of maximal oxygen consumption in a garter snake. *Am. J. Physiol.* **259,** R986–R992.
Garland, T., and Else, P. L. (1987). Seasonal, sexual and individual variation in endurance and activity metabolism in lizards. *Am. J. Physiol.* **252,** R439–R449.
Garland, T., Hankins, E., and Huey, R. B. (1990). Locomotor capacity and social dominance in male lizards (*Sceloporus occidentalis*). *Funct. Ecol.* **4,** 243–250.
Gatten, R. E. (1975). Effects of activity on blood oxygen saturation, lactate and pH in the turtles *Pseudemys scripta* and *Terrapene ornata*. *Physiol. Zool.* **48,** 24–35.
Gatten, R. E. (1981). Anaerobic metabolism in freely-diving painted turtles (*Chrysemys picta*). *J. Exp. Zool.* **216,** 377–385.
Gatten, R. E. (1985). The uses of anaerobiosis by amphibians and reptiles. *Am. Zool.* **25,** 945–965.
Gatten, R. E., Echternacht, A. C., and Wilson, M. A. (1988). Acclimatization versus acclimation of activity metabolism in a lizard. *Physiol. Zool.* **61,** 322–329.
Gauthier, J., Kluge, A. G., and Rowe, T. (1988a). The early evolution of the Amniota. *In* "The Phylogeny and Classification of the Tetrapods" (M. J. Benton, ed.), Vol. 1, pp. 103–155. Oxford Univ. Press (Clarendon), Oxford.
Gauthier, J., Kluge, A. G., and Rowe, T. (1988b). Amniote phylogeny and the importance of fossils. *Cladistics* **4,** 105–209.
Gleeson, T. T. (1979). Foraging and transport costs in the Galapagos marine iguana, *Amblyrhynchus cristatus*. *Physiol. Zool.* **52,** 549–557.
Gleeson, T. T. (1980). Lactic acid production during field activity in the Galapagos marine iguana, *Amblyrhynchus cristatus*. *Physiol. Zool.* **53,** 157–162.
Gleeson, T. T. (1981). Preferred body temperature, aerobic scope and activity capacity in the monitor lizard *Varanus salvator*. *Physiol. Zool.* **54,** 423–429.
Gleeson, T. T. (1991). Patterns of metabolic recovery from exercise in amphibians and reptiles. *J. Exp. Biol.* **160,** 187–207.
Gleeson, T. T., and Bennett, A. F. (1985). Respiratory and cardiovascular adjustments to exercise in reptiles. *In* "Circulation, Respiration and Metabolism: Current Comparative Approaches" (R. Gilles, ed.), pp. 23–38. Springer-Verlag, New York.
Gleeson, T. T., Mitchell, G. S., and Bennett, A. F. (1980). Cardiovascular responses

to graded activity in the lizards *Varanus* and *Iguana*. *Am. J. Physiol.* **239,** R174–R179.

Greene, H. W. (1988). Antipredator mechanisms in reptiles. *In* "Biology of the Reptilia" (C. Gans and R. B. Huey, eds.), Vol. 16, pp. 1–152. Wiley, New York.

Harvey, P. H., and Pagel, M. D. (1991). "The Comparative Method in Evolutionary Biology." Oxford Univ. Press, Oxford.

Hertz, P. E., Huey, R. B., and Nevo, E. (1982). Fight versus flight: Body temperature influences defensive responses of lizards. *Anim. Behav.* **30,** 676–679.

Hertz, P. E., Huey, R. B., and Garland, T. (1988). Time budgets, thermoregulation and maximal locomotor performance: Are reptiles Olympians or boy scouts? *Am. Zool.* **28,** 927–938.

Hochachka, P. W., and Somero, G. N. (1984). "Biochemical Adaptation." Princeton Univ. Press, Princeton, NJ.

Huey, R. B. (1987). Phylogeny, history and the comparative method. *In* "New Directions in Ecological Physiology" (M. E. Feder, A. F. Bennett, W. W. Burggren, and R. B. Huey, eds.), pp. 76–98. Cambridge Univ. Press, Cambridge, UK.

Huey, R. B., and Bennett, A. F. (1987). Phylogenetic studies of coadaptation: Preferred temperatures versus optimal performance temperatures of lizards. *Evolution (Lawrence, Kans.)* **41,** 1098–1115.

Huey, R. B., Pianka, E. R., and Schoener, T. W., eds. (1983). "Lizard Ecology: Studies of a Model Organism." Harvard Univ. Press, Cambridge, MA.

Huey, R. B., Bennett, A. F., John-Alder, H. B., and Nagy, K. A. (1984). Locomotor capacity and foraging behaviour of Kalahari lacertid lizards. *Anim. Behav.* **32,** 41–50.

Hutchison, V. H., Dowling, H. G., and Vinegar, A. (1966). Thermoregulation in a brooding female Indian python, *Python molurus bivittatus*. *Science* **151,** 694–696.

Jackson, D. C., and Prange, H. D. (1979). Ventilation and gas exchange during rest and exercise in adult green sea turtles. *J. Comp. Physiol.* **134,** 315–319.

Jayne, B. C. (1986). Kinematics of terrestrial snake locomotion. *Copeia*, pp. 915–925.

Jayne, B. C., and Bennett, A. F. (1990a). Scaling of speed and endurance in garter snakes: A comparison of cross-sectional and longitudinal allometries. *J. Zool.* **220,** 257–277.

Jayne, B. C., and Bennett, A. F. (1990b). Selection on locomotor performance capacity in a natural population of garter snakes. *Evolution (Lawrence, Kans.)* **44,** 1204–1229.

John-Alder, H. B., and Bennett, A. F. (1981). Thermal dependence of endurance, oxygen consumption and locomotory energetics in the lizard *Dipsosaurus dorsalis*. *Am. J. Physiol.* **241,** R342–R349.

John-Alder, H. B., Lowe, C. H., and Bennett, A. F. (1983). Thermal dependence of locomotory energetics and aerobic capacities of the gila monster (*Heloderma suspectum*). *J. Comp. Physiol.* **151,** 119–126.

John-Alder, H. B., Garland, T., and Bennett, A. F. (1986). Locomotory capacities, oxygen consumption and the cost of locomotion of the sleepy lizard (*Trachydosaurus rugosus* Gray). *Physiol. Zool.* **59,** 523–531.

Kram, R., and Taylor, C. R. (1990). Energetics of running—A new perspective. *Nature (London)* **346,** 265–267.

Lauder, G. V. (1981). Form and function: Structural analysis in evolutionary morphology. *Paleobiology* **7,** 430–442.

Lindstedt, S. L., and Jones, J. H. (1987). Symmorphosis: The concept of optimal design. *In* "New Directions in Ecological Physiology" (M. E. Feder, A. F. Bennett, W. W. Burggren, and R. B. Huey, eds.), pp. 289–305. Cambridge Univ. Press, Cambridge, UK.

Losos, J. B. (1990a). Ecomorphology, performance capability and scaling of West Indian *Anolis* lizards: An evolutionary analysis. *Ecol. Monogr.* **60,** 369–388.

Losos, J. B. (1990b). The evolution of form and function: Morphology and locomotor performance ability in West Indian *Anolis* lizards. *Evolution (Lawrence, Kans.)* **44,** 1189–1203.

Miles, D. B. (1989). Selective significance of locomotory performance in an iguanid lizard. *Am. Zool.* **29,** 146A.

Milstead, W. W., ed. (1967). "Lizard Ecology: A Symposium." Univ. of Missouri Press, Columbia.

Mitchell, G. S., Gleeson, T. T., and Bennett, A. F. (1981a). Ventilation and acid–base balance during graded activity in lizards. *Am. J. Physiol.* **240,** R29–R37.

Mitchell, G. S., Gleeson, T. T., and Bennett, A. F. (1981b). Pulmonary oxygen transport during activity in lizards. *Respir. Physiol.* **43,** 365–375.

Moberly, W. R. (1968a). The metabolic responses of the common iguana, *Iguana iguana*, to activity under restraint. *Comp. Biochem. Physiol.* **27,** 1–20.

Moberly, W. R. (1968b). The metabolic responses of the common iguana, *Iguana iguana*, to walking and diving. *Comp. Biochem. Physiol.* **27,** 21–32.

Nagy, K. A. (1982). Energy requirements of free-living iguanid lizards. *In* "Iguanas of the World: Their Behavior, Ecology and Conservation" (G. M. Burghardt and A. S. Rand, eds.), pp. 49–59. Noyes Publ., Park Ridge, NJ.

Nagy, K. A., Huey, R. B., and Bennett, A. F. (1984). Field energetics and foraging mode of Kalahari lacertid lizards. *Ecology* **65,** 588–596.

Pough, F. H. (1980). The advantages of ectothermy for tetrapods. *Am. Nat.* **115,** 92–112.

Pough, F. H., and Andrews, R. M. (1985a). Use of anaerobic metabolism by free-ranging lizards. *Physiol. Zool.* **58,** 205–213.

Pough, F. H., and Andrews, R. M. (1985b). Energy costs of subduing and swallowing prey for a lizard. *Ecology* **66,** 1525–1533.

Regal, P. J. (1978). Behavioral differences between reptiles and mammals: An analysis of activity and mental capabilities. *In* "Behavior and Neurology of Lizards: An Interdisciplinary Colloquium" (N. Greenberg and P. D. MacLean, eds.), pp. 183–202. U.S. Govt. Printing Office, Washington DC.

Ruben, J. A. (1976). Aerobic and anaerobic metabolism during activity in snakes. *J. Comp. Physiol.* **109,** 147–157.

Ruben, J. A. (1979). Blood physiology during activity in the snakes *Masticophis flagellum* (Colubridae) and *Crotalus viridis* (Crotalidae). *Comp. Biochem. Physiol. A* **64A,** 577–580.

Ruben, J. A., and Battalia, D. E. (1979). Aerobic and anaerobic metabolism during activity in small rodents. *J. Exp. Zool.* **208,** 73–76.

Ruben, J. A., and Bennett, A. F. (1980). The vertebrate pattern of activity metabolism: Its antiquity and possible relation to vertebrate origins. *Nature (London)* **286,** 886–888.

Ruben, J. A., and Bennett, A. F. (1981). Intense exercise and blood calcium levels in vertebrates. *Nature (London)* **291,** 411–413.

Schall, J. J., Bennett, A. F., and Putnam, R. W. (1982). Lizards infected with malaria: Physiological and behavioral consequences. *Science* **217,** 1057–1059.

Schmidt-Nielsen, K. (1972). Locomotion: Energy cost of swimming, flying and running. *Science* **177,** 222–228.

Secor, S. M., Jayne, B. C., and Bennett, A. F. (1992). Locomotor performance and energetic cost of sidewinding by the snake *Crotalus cerasates*. *J. Exp. Biol.* **163,** 1–14.

Seymour, R. S. (1982). Physiological adaptations to aquatic life. *In* "Biology of the Reptilia" (C. Gans and F. H. Pough, eds.), Vol. 13, pp. 1–51. Academic Press, London.

Taylor, C. R., and Weibel, E. R. (1981). Design of the mammalian respiratory system. I. Problem and strategy. *Respir. Physiol.* **44,** 1–10.

Taylor, C. R., Schmidt-Nielsen, K., and Raab, J. L. (1970). Scaling of energetic cost of running to body size in mammals. *Am. J. Physiol.* **219,** 1104–1107.

Tsuji, J. S., Huey, R. B., van Berkum, F. H., Garland, T., and Shaw, R. G. (1989). Locomotor performance of hatchling fence lizards (*Sceloporus occidentalis*): Quantitative genetics and morphometric correlates. *Evol. Ecol.* **3,** 240–252.

Tucker, V. A. (1967). The role of the cardiovascular system in oxygen transport and thermoregulation in lizards. *In* "Lizard Ecology: A Symposium" (W. W. Milstead, ed.), pp. 258–269. Univ. of Missouri Press, Columbia.

Ultsch, G. R., and Jackson, D. C. (1982). Long-term submergence at 3°C of the turtle, *Chrysemys picta belli*, in normoxic and severely hypoxic water. I. Survival, gas exchange and acid–base status. *J. Exp. Biol.* **96,** 11–28.

van Berkum, F. H., and Tsuji, J. S. (1987). Inter-familial differences in sprint speed of hatchling lizards (*Sceloporus occidentalis*). *J. Zool.* **212,** 511–519.

Vinegar, A., Hutchison, V. H., and Dowling, H. G. (1970). Metabolism, energetics and thermoregulation during brooding of snakes of the genus *Python* (Reptilia, Boidae). *Zoologica (N.Y.)* **55,** 19–48.

Vleck, D., Gleeson, T. T., and Bartholomew, G. A. (1981). Oxygen consumption during swimming in Galapagos marine iguanas and its ecological correlates. *J. Comp. Physiol.* **141,** 531–536.

Waldschmidt, S. R., Jones, S. M., and Porter, W. P. (1987). Reptilia. *In* "Animal Energetics" (T. J. Pandian and F. J. Vernberg, eds.), Vol. 2, pp. 553–619. Academic Press, San Diego.

Walton, M., Jayne, B. C., and Bennett, A. F. (1990). The energetic cost of limbless locomotion. *Science* **249,** 524–527.

Weibel, E. R., Taylor, C. R., and Hoppeler, H. (1991). The concept of symmorphosis: A testable hypothesis of structure function relationship. *Proc. Natl. Acad. Sci. U.S.A.* **88,** 10357–10361.

Weinstein, R. B., and Full, R. J. (1992). Intermittent exercise alters endurance in an eight-legged ectotherm. *Am. J. Physiol.* **262,** R852–R859.

Wilson, M. A., and Gatten, R. E. (1989). Aerobic and anaerobic metabolism of paired male lizards (*Anolis carolinensis*). *Physiol. Behav.* **46,** 977–982.

Wilson, M. A., Gatten, R. E., and Greenberg, N. (1990). Glycolysis in *Anolis carolinensis* during agonistic encounters. *Physiol. Behav.* **48,** 139–142.

Exercise Performance of Birds

DAVID. K. SAUNDERS[1] AND M. ROGER FEDDE

Department of Anatomy and Physiology, College of Veterinary Medicine, Kansas State University, Manhattan, Kansas 66506

I. Introduction
II. Energy Demands of Avian Exercise
  A. Flying
  B. Running
  C. Swimming and Diving
III. Physiological Adaptations for Prolonged Flights and Flights at High Altitude
  A. Oxygen Transport
  B. Thermoregulation and Water Conservation
IV. Physiological Adaptations for Diving
  A. Pulmonary Specializations
  B. Cardiovascular Specializations
  C. Tissue Specializations
V. Conditioning and Deconditioning in Birds
  A. Seasonal Changes in Muscle Mass
  B. Seasonal Changes in Muscle Biochemical Properties
VI. Summary
References

I. Introduction

Birds inhabit nearly all of the environments on earth, and their adaptations to these different environments have produced some remarkable and unique athletes. Most commonly, we associate birds with flight. Flight is a very energy-demanding form of exercise, as oxygen consumption of flying birds may increase by 7- to 14-fold over that at rest, a range similar to that found in many maximally exercis-

[1]Present Address: Division of Biological Sciences, Emporia State University, Emporia, KS 66801.

ing mammals. However, birds, unlike mammals, are capable of maintaining this high energy expenditure for long periods of time. Many migratory birds make long-distance migrations without stopping for food or water. For example, the Golden plover (*Pluvialis dominica*) migrates nonstop over water for about 3800 km between Hawaii and Alaska (Henshaw, 1910), a trip that takes about 37 h of continuous heavy exercise (Johnston and McFarlane, 1967).

Perhaps even more impressive is the ability of some avian species to migrate at altitudes where humans can barely survive, even at rest (Table I). The best known of these high-flying birds is the Bar-headed goose (*Anser indicus*), which flies at altitudes greater than 9000 m as it passes over the Himalayan peaks during its biannual migration between India and Tibet (Swan, 1970).

Not all birds are capable of flight, but even most of those that are not are excellent athletes. Penguins, for example, are effective flyers under water, reaching speeds of up to 32 km h^{-1}. Penguins also are

TABLE I

AVIAN SPECIES REPORTED TO FLY AT ALTITUDES ABOVE 6000 m[a]

Species	Altitude (m)
Andean condor[b]	6,035
Common crane[b]	6,096
Siberian white crane[c]	>6,096
Wall creeper[b]	6,400
Mallard duck[d]	6,401
Steppe eagle[e]	7,500
Lammergeier[b]	7,620
Common crane[c]	>7,620
Chough[f]	8,230
Bar-headed goose[f,g]	>8,848
Ruppell's griffon[h]	11,278

[a] Adapted from Table 1 of Fedde *et al.* (1985).

[b] Minertzhagen (1955).

[c] Walkinshaw (1973).

[d] Manville (1963).

[e] Brown (1976).

[f] Swan (1961).

[g] Swan (1970).

[h] Laybourne (1974).

capable of deep dives as they hunt for food. Emperor penguins (*Aptenodytes forsteri*) may reach depths greater than 250 m below the ocean surface (Kooyman *et al.*, 1971); furthermore, they return to the surface quickly, without suffering from nitrogen bubble formation in the blood. Terrestrial birds, such as the emu, ostrich, and road runner, are also excellent athletes. The ostrich can maintain running speeds of 40 km h^{-1} for more than 30 min, and can reach speeds of up to 70 km h^{-1} for shorter periods.

These are but a few examples of the remarkable athletic feats performed by birds. In this paper we will examine in detail some of those birds that are unique in their exercise capabilities and try to elucidate the physiological adaptations and mechanisms that allow these birds to survive and thrive in extreme environments.

II. Energy Demands of Avian Exercise

A. Flying

Flapping flight is a strenuous form of exercise, and oxygen consumption ($\dot{V}_{O_2}$) is an index of the energy demand. The maximal oxygen consumption ($\dot{V}_{O_2max}$) of mammals is related to body weight (M_b, in kg) by the equation

$$\dot{V}_{O_2max} = 40.4\ M_b^{0.87},$$

whereas for flying birds, maximal oxygen consumption is related to body weight by the equation

$$\dot{V}_{O_2max} = 48.6\ M_b^{0.69}$$

(Butler, 1991; Jones and Lindstedt, 1993). From these equations, small birds (<3 kg) would be predicted to have higher $\dot{V}_{O_2max}$ than mammals of similar size.

Measurements of oxygen consumption in flying birds range from about 5 to 14 times those measured at rest (Table II). The values for $\dot{V}_{O_2max}$ in mammals, including humans, indicate a similar aerobic scope. However, these magnitudes of oxygen consumption are not maximal for birds, because they can maintain flight for extended periods; maximal oxygen consumption has probably never been measured in a bird.

TABLE II

SPECIFIC OXYGEN CONSUMPTION ($\dot{V}_{O_2}$), SPECIFIC MINUTE VENTILATION ($\dot{V}_I$), SPECIFIC TIDAL VOLUME (V_T), AND RESPIRATORY FREQUENCY (f_{resp}) FOR SEVERAL AVIAN SPECIES AT REST AND DURING FLIGHT

	$\dot{V}_{O_2}$ (ml min^{-1} kg^{-1})			$\dot{V}_I$ (liters min^{-1} kg^{-1})			V_T (ml kg^{-1})			f_{resp} (breath min^{-1})		
Species	Rest	Flight	F/R	Rest	Flight	F/R	Rest	Flight	F/R	Rest	Flight	F/R
Pigeon[a]	—	—	—	0.118	2.452	20.7	4.64	5.0	1.08	26.0	487.0	18.7
Pigeon[b]	20.3	200.0	9.9	—	—	—	—	—	—	19.7	411.0	20.9
Budgerigar[c]	75.0	365.0	4.9	1.342	6.663	4.9	—	—	—	—	—	—
Fish crow[d]	30.4	260.0	8.6	0.796	6.390	8.0	8.20	14.9	1.80	27.3	120.0	4.4
Evening grosbeak[e]	40.5	568.3	14.0	1.350	17.030	12.6	0.85	3.5	4.10	95.0	294.0	3.1
Black duck[e]	18.2	231.0	12.7	0.611	8.390	13.7	22.50	53.0	2.35	27.2	158.0	5.8
Starling[f]	37.18	360.3	9.7	0.782	6.460	8.3	0.67	2.8	4.18	92.0	180.0	2.0
Kestrel[g]	29.2	364.6	12.5	—	—	—	—	—	—	—	—	—
Barnacle goose[h]	—	—	—	—	—	—	—	—	—	8.5	99.0	11.6
Mallard[i]	—	—	—	—	—	—	—	—	—	14.0	96.0	6.9

[a] Hart and Roy (1966).
[b] Butler *et al.* (1977).
[c] Tucker (1968b). Resting data for Budgerigar from Greenwald *et al.* (1967).
[d] Bernstein (1976).
[e] Berger *et al.* (1970).
[f] Torre-Bueno (1976, 1978a).
[g] Gessaman (1980).
[h] Butler and Woakes (1980).
[i] Lord *et al.* (1962).

B. Running

Studies of naturally cursorial birds running on a treadmill have shown that oxygen consumptions of these birds are nearly equal to those recorded for flying birds. Ostriches running on a treadmill at speeds lower than they run in their natural environment are capable of increasing oxygen consumption by about 10-fold above that at rest (Fedak and Seeherman, 1979). Grubb *et al.* (1983) found oxygen consumption of emus running at the experiment's highest exercise intensity to be 11.4 times that at rest. The authors stated that even at the highest exercise intensities the birds did not appear stressed, and were likely capable of running faster and increasing their oxygen consumption further.

C. Swimming and Diving

Most studies of metabolic rate in diving birds have been performed under artificial conditions in the laboratory. These studies have resulted in wide variation in the measurement of oxygen consumption of several different species. The metabolic rate of Little Blue penguins swimming underwater at 0.85 m s^{-1} increases only 1.3 times that at rest (Baudinette and Gill, 1985). Similar results were found for the Humboldt penguin, with oxygen consumption increasing by only 26% during diving (Butler and Woakes, 1984), and for King penguins, in which metabolic rate increases by only 2.8 times the standard metabolic rate during foraging (Kooyman *et al.*, 1982). In contrast, the metabolic rate of Jackass penguins increases 10 times above the resting metabolic rate during hunting and diving (Nagy *et al.*, 1984).

Swimming at the surface of the water appears to be more energy demanding for birds than does swimming under water. Mallard ducks swimming at maximum sustainable speeds increase their oxygen consumption by more than fourfold above that at rest (Prange and Schmidt-Nielsen, 1970), and Tufted ducks swimming at 0.71 m s^{-1} increase their oxygen consumption by more than twofold (Butler *et al.*, 1988). Nevertheless, both swimming and diving appear to be less demanding forms of exercise than flapping flight. This likely is the result of the muscle mass involved in these forms of exercise, as the ratio of flight muscle to leg muscle in the Mallard duck is 2.3:1 (Prange and Schmidt-Nielsen, 1970). Thus, a smaller muscle mass is used for swimming compared to flight, resulting in a lower maximum metabolic rate during swimming. Further, the metabolic capacity of the flight muscles is much greater than that of the leg muscles in birds that normally fly (Butler, 1991).

III. Physiological Adaptations for Prolonged Flights and Flights at High Altitude

Despite the high energy cost of flight, many birds cover great distances while migrating, and do so without resting or stopping for food or water (Odum *et al.*, 1961, Tucker, 1971). Aerobic metabolism is likely used during prolonged flights and flights performed at high altitude, as birds cannot continuously accumulate lactate during these flights, as would occur if anaerobic metabolism were used. Also associated with the high energy cost of flight is an increased body temperature; thus, birds must have adaptations that enable them to remove excess heat during long flights. In addition, because they do not stop for water, water loss is also a concern for a migrating bird. Birds possess many unique adaptations that allow them to compensate for these problems.

A. Oxygen Transport

The capabilities and efficacy of the oxygen transport system of birds are best characterized by those high-altitude-adapted birds that are capable of maintaining aerobic metabolism in severely hypoxic environments. The oxygen transport system consists of three basic components: (A) the pulmonary system; (B) the cardiovascular system; and (C) the tissues or body cells. We will investigate the adaptations within each of these components that enhance the ability of birds to perform rigorous exercise for prolonged periods, even in extremely hypoxic environments.

1. Pulmonary Specializations

a. The Crosscurrent Gas Exchanger. The anatomy of the respiratory system in birds differs from that of other vertebrates. Functional evidence suggests that in the avian lung the flow of blood through the capillaries is at right angles to the convective flow of gas through the lung. It is this anatomical relationship between blood flow and gas flow that establishes the crosscurrent gas exchange system of the avian lung.

The crosscurrent gas exchange system provides a greater gas exchange efficacy compared to a uniform pool system, for example, a mammalian lung (Piiper and Scheid, 1975). This increased efficacy, however, appears to occur only during exposure to nonhypoxic environments. A comparison of the avian parabronchial lung to the mammalian alveolar lung over a range of inspired oxygen partial pressures

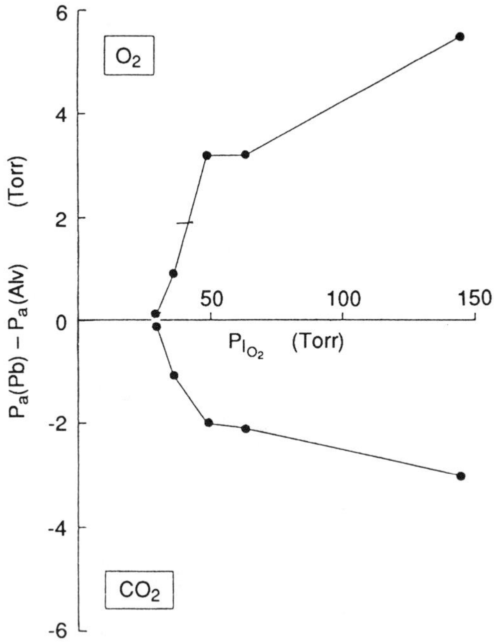

FIG. 1. Advantage of parabronchial compared with alveolar gas exchange. Arterial P_{O_2} (upper) and P_{CO_2} (lower) values measured in the bird were recalculated assuming an alveolar-type gas exchange system, but leaving all other pertinent variables (e.g., ventilation, diffusing capacity, cardiac output) unaltered. The difference between arterial partial pressures in the parabronchial lung, P_a(Pb), and in the alveolar lung, P_a(Alv), reflects the advantage of the higher gas exchange efficacy of the avian, compared to mammalian, lung. From Shams and Scheid (1989); reprinted with permission.

was performed by Shams and Scheid (1989). The parabronchial lung yielded higher arterial P_{O_2} than the mammalian lung at all inspired partial pressures; however, the difference decreased with decreasing inspired P_{O_2} (Fig. 1). As a result of the decreased difference in arterial P_{O_2} between birds and mammals at low inspired oxygen concentration, Shams and Scheid (1989) suggested that the advantage of crosscurrent gas exchange in the avian lung is negligible under conditions of deep hypoxia, and thus at extreme altitudes.

The structural design of the avian lung appears to be more advantageous for gas exchange than is that of the mammalian lung. Morphometric studies have shown the lungs of birds to have a thinner blood–gas diffusion barrier than that of mammals. This diffusion barrier is usually expressed morphometrically as the mean harmonic

TABLE III

COMPARISON OF HARMONIC MEAN THICKNESS OF THE BLOOD–GAS BARRIER IN SELECTED SPECIES OF BIRDS AND MAMMALS

Avaian species	Mean harmonic thickness (μm)	Mammalian species	Mean harmonic thickness (μm)
African rock martin[a]	0.090	Pygmy shrew[g]	0.260
House sparrow[a]	0.096	Eurasian water shrew	0.290
Violet-eared hummingbird[b]	0.099	Savi's pygmy shrew	0.270
House sparrow[b]	0.118	Waltzing mouse	0.260
Budgerigar[b]	0.118	White mouse	0.290
Graylag goose[c]	0.118	European white-toothed shrew	0.360
Grosbeak weaver[a]	0.120	White rat	0.370
Singing cisticola[a]	0.122	Thompson's gazelle	0.370
Olive thrush[a]	0.1270	Wildebeest	0.373
Mallard[c]	0.133	African giant shrew	0.380
Common starling[a]	0.141	Banded mongoose	0.409
Baglafecht weaver[a]	0.151	Guinea pig	0.420
Fiscal shrike[a]	0.170	Dark forest shrew	0.420
Domestic pigeon[d]	0.172	Dik-dik	0.429
Domestic goose[e]	0.260	Waterbuck	0.456
Canada goose[e]	0.290–0.300	Dog	0.483[h]

Chicken[f]	0.314	Rabbit	0.500
Domestic fowl[d]	0.346	Eland	0.500
Turkey[d]	0.385	Zebu cattle	0.500
		Monkey	0.500
		Genet cat	0.506
		Swiss cow	0.510
		African sheep	0.530
		Grants gazelle	0.537
		Suni	0.562
		Camel	0.600
		Horse	0.600
		Giraffe	0.600
		Human	0.620
Mean	0.175	Mean	0.444

[a] Maina (1984).
[b] Dubach (1981).
[c] Maina and King (1982).
[d] Duncker (1973).
[e] Powell and Mazzone (1983).
[f] Abdalla *et al.* (1982).
[g] Data for all mammalian species are from Gehr *et al.* (1981).
[h] Mean value.

TABLE IV

SPECIFIC OXYGEN CONSUMPTION, SPECIFIC MINUTE VENTILATION, SPECIFIC TIDAL VOLUME, AND RESPIRATORY FREQUENCY OF VARIOUS BIRDS AT REST AND DURING EXERCISE (EX) UNDER NORMOXIC AND HYPOXIC CONDITIONS

	Specific oxygen consumption ($ml\ min^{-1}\ kg^{-1}$)						Specific minute ventilation ($liters\ min^{-1}$)					
	Normoxia			Hypoxia			Normoxia			Hypoxia		
Species	Rest	Ex	Ex/Rest	Rest	Ex	Ex/Rest	Rest	Ex	Ex/Rest	Rest	Ex	Ex/Rest
Bar-headed goose[a]	20.2	44.8	2.2	15.7	22.4	1.4	0.34	1.57	4.6	0.71	2.48	3.5
Bar-headed goose[b]	12.6	—	—	21.6	—	—	0.29	—	—	2.44	—	—
Pekin duck[c]	12.1	36.0	3.0	10.7	38.7	3.6	0.32	1.25	3.9	0.49	2.12	4.3
Pekin duck[b]	18.9	—	—	20.2	—	—	0.42	—	—	1.84	—	—
Pekin duck[d]	14.8	—	—	10.2	—	—	0.35	—	—	2.01	—	—
House sparrow[e]	75.3	—	—	92.3	—	—	1.34	—	—	2.37	—	—

[a] Fedde *et al.* (1989).
[b] Black and Tenney (1980a).
[c] Kiley *et al.* (1985).
[d] Shams and Scheid (1989).
[e] Tucker (1968a).

thickness of the blood–gas barrier and is representative of the tissue resistance to diffusion (Weibel, 1972). Values for the mean harmonic thickness of the blood–gas barrier of the lungs of various birds and mammals are given in Table III.

Not only do birds have a thinner blood–gas barrier than mammals, they also have exchange surfaces that are two to four times larger than those of mammals, based on the volumes of respiratory tissue (Dubach, 1981). Both the decrease in blood–gas barrier thickness and increased exchange surface area of the avian lung lead to much higher diffusing capacities than in the mammalian lung (Dubach, 1981). An increased diffusing capacity would be an advantage for the high oxygen consumption requirements of flight, as well as during exposure to hypoxia.

b. Ventilation. Ventilation increases dramatically with the onset of flight and seems to be well matched with oxygen consumption for most birds studied (Table II). However, hyperventilation has been shown in flying pigeons and starlings, in which decreases in arterial and clavicular air sac P_{CO_2} were found (Butler *et al.*, 1977; Torre-Bueno, 1978a). Thus, ventilation does not appear to be a limiting factor for oxygen consumption during flight under normoxic conditions.

Ventilation also increases in resting birds exposed to hypoxic envi-

TABLE IV (*Continued*)

Specific tidal volume (ml kg^{-1})						Respiratory frequency (breath min^{-1})					
Normoxia			Hypoxia			Normoxia			Hypoxia		
Rest	Ex	Ex/Rest	Rest	Ex	Ex/Rest	Rest	Ex	Ex/Rest	Rest	Ex	Ex/Rest
28.3	31.8	1.1	40.5	44.5	1.1	11.8	47.9	4.1	17.6	55.5	3.2
—	—	—	—	—	—	—	—	—	—	—	—
29.2	16.3	3.0	24.6	25.8	1.1	11.0	74.0	6.7	18.0	79.0	4.4
—	—	—	—	—	—	—	—	—	—	—	—
23.5	—	—	44.4	—	—	15.5	—	—	46.6	—	—
16.5	—	—	21.1	—	—	81.0	—	—	112.0	—	—

ronments compared to resting birds in normoxic environments. (Table IV). Further increases in ventilation occur when noncursorial birds are forced to run under hypoxic conditions (Table IV). This increase in ventilation appears to be adequate to meet the oxygen demands of the working tissues. To date, few studies have been conducted to determine the effect of flight in a hypoxic environment on ventilation, or to determine if ventilation is a limiting factor for flight in hypoxia.

The increased ventilation associated with flight in normoxia results from increases in both respiratory frequency and tidal volume, whereas in exercise under hypoxic conditions, ventilation increases primarily due to increased respiratory frequency (Tables II and IV). Increased ventilation benefits gas exchange by increasing the amount of gas exposed to the gas exchange surface, decreasing P_{CO_2}, and increasing P_{O_2} in the gas exchanger (Fedde *et al.*, 1985). It is unknown whether birds flying in hypoxia can increase their ventilation above that during flight in normoxia, or whether such increases would be accomplished by increases in tidal volume, respiratory frequency, or both.

Increased ventilation also has been shown during flight at high ambient temperatures in the Fish crow and White-necked raven (Bernstein, 1976; Hudson and Bernstein, 1981). Indeed, hyperventilation during flight has been suggested to be caused by hyperthermia, as even at ambient temperature of 0°C, body temperature increases by about 2°C during flight (Torre-Bueno, 1976). However, Kiley *et al.* (1982) found that hyperventilation was not prevented in Pekin ducks during running exercise when the ducks were kept at constant body temperature.

If hyperventilation is not the result of increased body temperature, then what causes it? Gleeson and Brackenbury (1983) suggested that hyperventilation in running isothermic ducks may be caused by metabolic acidosis. This may well be the case for Pekin ducks running on a treadmill. However, the *pectoralis* muscles of birds are vastly different from the leg muscles in terms of factors involved in the movement of oxygen to these tissues (e.g., capillary density, fiber type, enzymatic properties). Thus, during treadmill exercise, delivery of oxygen to the leg muscles may not be adequate and net anaerobic metabolism would be used, resulting in the accumulation of lactic acid. If hyperventilation occurs during long-distance flights, accumulation of lactic acid is not likely to be the stimulus because, as suggested earlier, these flights presumably must be performed using entirely aerobic power.

Birds, unlike mammals, lack a functional diaphragm and depend on movement of the sternum for ventilation. Because the primary flight muscles (*pectoralis* and *supracoracoideus*) originate from the sternum, it seems plausible that movement of the wings might also influence ventilation during flight. If this were the case, ventilatory frequency would be expected to be synchronized in a 1:1 ratio with wing beat frequency. However, to date, only two species studied have shown such synchronicity. Many birds during flight have a constant ratio of wing beats to breaths (2:1, 3:1, etc.), referred to as coordination, whereas others have no apparent synchronicity or coordination (Table V).

The exact function of the wing beat/ventilatory frequency coordination that is apparent for some birds is unknown. No apparent obligatory link exists between ventilation and wing beat frequency, as shifts in the coordination ratio can change during flight (Berger *et al.*, 1970). Further, Tucker (1969) found that Budgerigars had constant wing beat frequencies, but as flight speed increased, so did ventilation rate. Bernstein (1987) has suggested that the coordination of wing beat frequency to ventilation rate may be part of a complex relationship between ventilation and perfusion of the lung during flight, as cardiac motion also appears to be coordinated with wing beat frequency (Aulie, 1971, 1972). However, Banzett *et al.* (1992) recorded respiratory flow and the timing of wing beats in the Starling. They found that only a small fraction (3 to 11%) of the tidal volume was associated with the wing beat. Even this small fraction was thought to be an overestimate by the authors. Thus, they concluded that wing beat and breathing were not mechanically linked in this bird.

c. Conversion to an Ideal Lung. The avian lung becomes a near-perfect gas exchange organ when birds at rest are exposed to hypoxia.

TABLE V

RATIO OF WING BEAT FREQUENCY (f_{wb}) TO RESPIRATORY FREQUENCY (f_{resp}) IN SEVERAL SPECIES OF BIRDS

Species	$f_{wb} : f_{resp}$
Crow[a]	1.0 : 1
Pigeon[a]	1.0 : 1
Mallard duck[b]	2.0 : 1
Western gull[c]	3.0 : 1
Barnacle goose[d]	3.0 : 1
Starling[e]	3.0 : 1
Pintail[c]	4.6 : 1
Lesser scaup[c]	4.9 : 1
Black duck[a]	5.0 : 1
Quail[a]	5.0 : 1
Pheasant[a]	5.0 : 1
Mallard duck[c]	5.3 : 1
Black-rumped waxbill[a]	Not coordinated
Black-capped chickadee[a]	Not coordinated

[a] Berger *et al.* (1970).
[b] Lord *et al.* (1962).
[c] Tomlinson (1963).
[d] Butler and Woakes (1980).
[e] Banzett *et al.* (1992).

As the partial pressure of the moist inspired gas becomes progressively lower, the difference between the inspired partial pressure of oxygen ($P\text{I}_{O_2}$) and the arterial oxygen partial pressure ($P\text{a}_{O_2}$) decreases (Tucker, 1968a; Black and Tenney, 1980a; Faraci *et al.*, 1984a; Kiley *et al.*, 1985; Fedde *et al.*, 1989). The difference between $P\text{I}_{O_2}$ and $P\text{a}_{O_2}$ seen under hypoxic conditions at rest can also be reduced when birds exercise under these conditions (Kiley *et al.*, 1985; Fedde *et al.*, 1989). Bar-headed geese exposed to $P\text{I}_{O_2}$ of 48 Torr had $P\text{a}_{O_2}$ of 31 Torr at rest. During running exercise at the same $P\text{I}_{O_2}$, $P\text{a}_{O_2}$ increased to 37 Torr (Fedde *et al.*, 1989). Because these $P\text{a}_{O_2}$ changes occur on the steep slope of the oxyhemoglobin equilibrium curve of these birds, a small increase in $P\text{a}_{O_2}$ can greatly increase the arterial oxygen concentration. For example, the 6-Torr increase in $P\text{a}_{O_2}$ of exercising Bar-headed geese resulted in a 32% increase in the concentration of oxygen in the arterial blood (Fedde *et al.*, 1989).

The mechanisms by which the avian lung can become an ideal gas exchanger under hypoxic conditions have yet to be completely elucidated. It has been suggested that the avian lung is capable of extremely high ventilation rates when exposed to hypoxic conditions

(Shams and Scheid, 1987, 1989). This would result in complete lung gas mixing and increase the amount of gas exposed to the gas exchange surface. However, this ability for high ventilation is not unique to birds. Human high-altitude natives and sojourners climbing Mount Everest are also capable of extremely high ventilations. The fractional increase in ventilation above that at control (ambient) P_{O_2} is nearly equal for humans and Bar-headed geese over a range of decreasing P_{O_2} (Monge and Leon-Velarde, 1991). However, the conversion to an ideal lung in birds does not appear to solely result from improvement in the matching of the ventilation/perfusion conductance ratio (Powell, 1982).

It is interesting to note that the avian lung apparently does not become an ideal gas exchanger during normoxic flight or during other forms of exercise under normoxic conditions. Oxygen exchange also appears to be unaffected in birds running on a treadmill breathing normoxic gas mixtures, as the oxygen concentration and partial pressure in the arterial blood are unchanged from rest to exercise (Bech and Nomoto, 1982; Grubb *et al.*, 1983). However, Butler *et al.* (1977) found a small decrease in the difference between PI_{O_2} and Pa_{O_2} in flying pigeons.

Exposure to cold ambient temperatures may increase parabronchial oxygen extraction. Exposure of the European Coot (*Fulica atra*) to ambient temperature of −25°C resulted in a 60% increase in ventilation over that at 20°C, whereas oxygen consumption increased 350% (Brent *et al.*, 1984). Therefore, parabronchial oxygen extraction was more than doubled at −25°C compared to at 20°C. Increases in oxygen extraction with exposure to cold temperatures have also been shown in the Linneated parakeet (*Bolborhynchos lineola*), Mallard duck (*Anas platyrhynchos*), Little penguin (*Eudyptula minor*), and the Kittiwake (*Rissa tridactyla*) (Bucher, 1981; Brent *et al.*, 1983; Bech *et al.*, 1985; Johannesen and Nicol, 1990). The ability to increase oxygen extraction with exposure to cold temperatures could be of value to high-altitude-flying Bar-headed geese that are exposed to extreme hypoxia and cold temperatures during their flight over the Himalayas.

The mechanism for increased parabronchial oxygen extraction with relatively small increase in ventilation during cold exposure is currently unknown, but may result from better matching of ventilation to perfusion. Powell and Wagner (1982) found unequal ventilation to perfusion conductance ratios in the lung units of geese, in which a lung unit could receive as much as approximately 10% of the total ventilation while receiving as little as 0.3% of the cardiac output. Thus, Bech *et al.* (1984) suggested that decreased ventilation in cold-exposed birds could enhance the ventilation to perfusion conductance ratio.

TABLE VI

ALLOMETRIC EQUATIONS AND PREDICTED VALUES FOR HEART MASS, RESTING HEART RATE, RESTING STROKE VOLUME, AND RESTING CARDIAC OUTPUT OF BIRDS AND MAMMALS AT REST OVER A RANGE OF BODY MASS (M_b, IN KG)

	Heart mass (M_b, g)			Heart rate (f_b, min^{-1})			Stroke volume (V_s, ml)			Cardiac output ($\dot{Q}$, ml min^{-1})		
	Bird[a]	Mammal[b]	Bird/Mammal	Bird[a]	Mammal[c]	Bird/Mammal	Bird[a]	Mammal[c]	Bird/Mammal	Bird[a]	Mammal[c]	Bird/Mammal
Allometric equation	$M_h = 8.76\ M_b^{0.92}$		$M_h = 5.8\ M_b^{0.98}$	$f_h = 179\ M_b^{-0.28}$		$f_h = M_b^{-0.25}$	$V_s = 1.72\ M_b^{0.97}$		$V_s = 0.66\ M_b^{1.05}$	$\dot{Q} = 291\ M_b^{0.69}$		$\dot{Q} = 166\ M_b^{0.79}$
Body mass (kg)												
0.01	0.13	0.06	1.98	654	746	0.88	0.02	0.005	4.00	12.1	4.4	2.77
0.1	1.05	0.61	1.73	342	420	0.81	0.18	0.06	3.12	59.4	26.9	2.20
1	8.76	5.80	1.61	179	236	0.76	1.72	0.66	2.61	291	166	1.75
10	72.9	55.4	1.32	93.3	133	0.70	16.1	7.41	2.17	1424	1024	1.39
100	606	529	1.15	48.7	74.6	0.65	150	83.1	1.80	6973	6311	1.10

[a] Grubb (1983).
[b] Stahl (1967).
[c] Holt *et al.* (1968).

Increased oxygen extraction has not been shown in all avian species during exposure to cold temperatures, as demonstrated by pigeons (*Columba livia*) (Bouverot *et al.*, 1976; Barnas and Rautenberg, 1984; Bech *et al.*, 1985). Bech *et al.* (1985) suggest that species differences apparently exist for the ability to increase oxygen extraction during cold exposure. These differences between species have yet be elucidated, however.

2. *Cardiovascular Specializations*

a. Cardiac Variables. Allometric relationships suggest that the avian cardiovascular system is better adapted for convective transport of oxygen, and thus for high-intensity exercise, than is that of mammals. Birds at rest typically have a larger heart mass, stroke volume, and cardiac output than do mammals of the same size (Table VI). The larger cardiac outputs of birds at rest result from increased stroke volume, as resting birds have lower heart rates than do resting mammals of similar size (Table VI).

The allometric equations for birds and mammals for heart mass, cardiac output, stroke volume, and heart rate differ not only in coefficients, but also in their mass exponents. For each of these variables, as well as for $\dot{V}_{O_2max}$, the mass exponent is lower for birds than for mammals. Therefore, birds weighing less than 1 kg have higher values for heart mass, stroke volume, and cardiac output than do mammals of equal weight (Table VI), resulting in cardiac output being nearly three times greater in a 10-g bird than in a 10-g mammal. The advantage, if any, of the increased cardiac output for small birds at rest compared to resting small mammals is not known. Interestingly, as the sizes of birds and mammals increase, the mass-specific cardiac outputs of birds and mammals become nearly equal (Table VI).

All of these allometric relationships compare cardiovascular variables between birds and mammals at rest. How do these variables compare during exercise? Butler *et al.* (1977) found that pigeons flying in a wind tunnel increased cardiac output more than 4-fold above resting values (Table VII). Increased cardiac output was primarily due to increased heart rate. Similar increases in heart rate during flight have been shown for numerous other birds (Table VII). Because of technical difficulties with obtaining cardiac output measurements from flying birds, most work has been performed on birds running on a treadmill. Grubb (1983) found a 7.3-fold increase in cardiac output during exercise compared to rest in the emu (Table VII). This increase resulted from a 3.9-fold increase in heart rate and a 1.8-fold increase in stroke

TABLE VII

SPECIFIC CARDIAC OUTPUT, HEART RATE, AND SPECIFIC STROKE VOLUME FOR VARIOUS BIRDS DURING EXERCISE (EX) IN FLIGHT AND TREADMILL RUNNING OR WALKING

	Cardiac output (liters min^{-1} kg^{-1})			Heart rate (beat min^{-1})			Stroke volume (ml kg^{-1})		
Species	Rest	Ex	E/R	Rest	Ex	E/R	Rest	Ex	E/R
				Flight					
Pigeon[a]	0.555	2.41	4.38	171	530	3.10	3.26	3.63	1.10
Pigeon[b]	—	—	—	168	558	3.32	—	—	—
Black-capped chickadee[b]	—	—	—	618	1038	1.68	—	—	—
Chipping sparrow[b]	—	—	—	498	981	1.97	—	—	—
House sparrow[b]	—	—	—	534	951	1.78	—	—	—
Wood duck[b]	—	—	—	156	588	3.77	—	—	—
Barnacle goose[c]	—	—	—	173	530	3.06	—	—	—
Kestrel[d]	—	—	—	312	710	2.28	—	—	—
				Treadmill Exercise					
Emu[e]	0.067	0.494	7.28	45.8	180	3.93	1.52	2.74	1.80
Pigeon[f]	0.341	0.700	2.05	178	260	1.46	1.97	2.70	1.37
Bar-headed goose[g]	0.240	0.460	1.92	250	420	1.68	0.97	1.10	1.13
Pekin duck[f]	0.304	0.700	2.30	199	440	2.21	1.60	1.60	1.00
Pekin duck[h]	0.310	0.536	1.73	301	501	1.66	1.08	1.08	1.00
Pekin duck[i]	0.540	0.870	1.61	174	328	1.88	3.13	2.77	0.88
Greyhound[j]	0.214	0.914	4.27	97.0	318	3.28	2.20	2.90	1.32
Human[k]	0.070	0.429	6.00	50.0	200	4.00	1.43	2.14	1.50

[a] Butler *et al.* (1977).
[b] Berger *et al.* (1970).
[c] Butler and Woakes (1980).
[d] Gessaman (1980).
[e] Grubb *et al.* (1983).
[f] Grubb (1982).
[g] Fedde *et al.* (1989).
[h] Kiley *et al.* (1985).
[i] Bech and Nomoto (1982).
[j] Staaden (1984).
[k] Åstrand *et al.* (1964).

volume. Running birds that are not naturally cursorial increase cardiac output much less than the emu (Table VII), and pigeons running on a treadmill have much lower cardiac output than do pigeons during flight (Table VII). Thus, treadmill exercise probably is not a good indicator of maximum cardiac output in birds that normally fly.

Cardiac output changes more for a bird than for a mammal for a given increase in oxygen consumption. In Pekin ducks, pigeons, and emus running on a treadmill, cardiac output increases between 9.5 and 10.2 ml for each 1 ml increase in oxygen consumption (Grubb, 1982; Grubb *et al.*, 1983). These are greater than the changes reported for dogs, humans, and horses, for which cardiac output increases between 6 and 7 ml for each 1 ml increase in oxygen consumption (Barger *et al.*, 1956). The increased cardiac output in exercising birds could allow a greater quantity of oxygen to be delivered to the tissues per unit time.

Cardiac output also increases in birds exposed to hypoxia; however, differences exist in the cardiac output responses of high- and low-altitude-adapted birds. Low-altitude-adapted Pekin ducks increase by two- and fourfold at $P\text{I}_{O_2}$ of 37 and 27 Torr, respectively, compared to their cardiac output at $P\text{I}_{O_2}$ of 146 Torr (Black and Tenney, 1980a). In contrast, high-altitude-adapted Bar-headed geese had cardiac outputs that were nearly equal at both 146 and 37 Torr, yet at $P\text{I}_{O_2}$ of 27 Torr, cardiac output increased by sevenfold (Black and Tenney, 1980a). Thus, it appears that Bar-headed geese have a greater cardiac output reserve capacity to be utilized during hypoxia than do Pekin ducks.

Despite the adaptations for high cardiac output in birds, cardiac output may be a major factor in limiting the altitude at which birds fly. Cardiac output initially increases with decreasing $P\text{I}_{O_2}$; however, a $P\text{I}_{O_2}$ is eventually reached where cardiac output begins to decrease (Black and Tenney, 1980a; Shams and Scheid, 1989). Bar-headed geese running in hypoxia have cardiac output that is nearly half that when exercising in normoxia (Table VIII). Thus, it appears that maximal cardiac output may be limited by an inhibitory effect of hypoxia on the heart itself and thus may limit the altitudes at which birds fly (Fedde *et al.*, 1989; Shams and Scheid, 1989). How the heart is capable of meeting the demands of flight under the extreme hypoxic conditions at high altitude remains to be determined.

b. Arterial Blood Pressure. Birds typically have higher mean arterial blood pressure than mammals (Grubb, 1983). The effect of exercise on mean arterial blood pressure seems to be species specific. Arterial blood pressure rises significantly with treadmill exercise in the Pekin duck and Turkey (Baudinette *et al.*, 1982; Bech and Nomoto, 1982; Grubb, 1982) but is unaffected in the Bar-headed goose, emu, and

TABLE VIII

SPECIFIC CARDIAC OUTPUT, SPECIFIC STROKE VOLUME, AND HEART RATE IN THE PEKIN DUCK AND BAR-HEADED GOOSE AT REST DURING EXERCISE (EX) UNDER NORMOXIC AND HYPOXIC CONDITIONS

	Specific cardiac output ($ml\ min^{-1}\ kg^{-1}$)						Specific stroke volume ($ml\ kg^{-1}$)						Heart rate ($beat\ min^{-1}$)					
	Normoxia			Hypoxia			Normoxia			Hypoxia			Normoxia			Hypoxia		
Species	Rest	Exercise	Ex/Rest	Rest	Exercise	Ex/Rest	Rest	Exercise	Ex/Rest	Rest	Exercise	Ex/Rest	Rest	Exercise	Ex/Rest	Rest	Exercise	Ex/Rest
Pekin duck[a]	0.31	0.54	1.74	0.28	0.60	2.14	1.08	1.08	1.00	1.04	1.17	1.13	301	501	1.66	290	519	1.79
Bar-headed goose[b]	0.24	0.46	1.92	0.22	0.25	1.14	0.96	1.08	1.13	0.79	0.58	0.73	250	460	1.84	280	430	1.54

[a] Kiley *et al.* (1985).
[b] Fedde *et al.* (1989).

pigeon (Grubb, 1982; Grubb *et al.*, 1983; Fedde *et al.*, 1989). Further, mean arterial blood pressure is not significantly changed from rest during flight in the pigeon (Butler *et al.*, 1977).

Mean arterial blood pressure is not significantly affected in most birds exposed to hypoxia (Faraci *et al.*, 1984b; Faraci and Fedde, 1986; Fedde *et al.*, 1989). However, mean arterial blood pressure in Bar-headed geese increases above that at rest during treadmill exercise under hypoxic conditions (Fedde *et al.*, 1989).

The pulmonary arterial pressure response to hypoxia also appears to be species specific in birds. Pulmonary arterial pressure increases in Pekin ducks exposed to hypoxia, whereas no increase occurs in Bar-headed geese, even during severe hypoxia (Black and Tenney, 1980b; Faraci *et al.*, 1984b). Bar-headed geese might be less likely to develop right heart hypertrophy and pulmonary edema when exposed to hypoxia as a result of their blunted pulmonary arterial pressure response (Faraci, 1988).

c. Blood Properties. In general, there is little difference in hematocrits and hemoglobin concentrations of birds and mammals (Lutz *et al.*, 1974). The oxygen partial pressures at which hemoglobin is 50% saturated (P_{50}) appear to be higher in avian blood compared to mammalian blood (Table IX). However, when the P_{50} of avian and mammalian bloods are compared mass specifically, little difference is seen between mammals and birds of similar size (Lutz *et al.*, 1974; Table X).

Both high-altitude-adapted birds and mammals tend to have hemoglobins with high affinity for oxygen and low P_{50} (Table IX). A comparison of the Bar-headed goose (P_{50} = 27.2 Torr) to the Pekin duck (P_{50} = 42.6 Torr) serves to illustrate the advantage conferred by high-oxygen-affinity hemoglobin at high altitude. At identical Pa_{O_2} (28 Torr) and similar hematocrits, a unit of goose blood can carry 2.5 times more oxygen than a unit of duck blood. Thus, oxygen delivery in the goose can equal that in the duck with much lower blood flow, or, for equal blood flow, the goose can greatly enhance oxygen delivery to the tissues (Faraci *et al.*, 1984a). Nevertheless, systemic oxygen transport capacity does not necessarily predict oxygen delivery to the tissues from the capillaries (Sarelius, 1989).

The oxygen affinity of avian blood increases with decreases in Pa_{CO_2} and increases in pH. Arterial pH has been shown to increase as high as 7.7 during exposure to severe hypoxia in resting Bar-headed geese (Fedde *et al.*, 1985). Using Black and Tenney's (1980a) regression equation describing the Bohr effect for the blood of Bar-headed geese ($\log P_{50} = 5.41 - [0.53 \times \text{pH}]$) and a pH of 7.7, the P_{50} is estimated to be 21.3 Torr. During flight under hypoxic conditions, it is likely that ventilation would increase to higher values to enhance oxygen uptake by

TABLE IX

HALF-SATURATION PRESSURES (P_{50}, IN TORR) FOR THE BLOOD OF VARIOUS BIRDS AND MAMMALS

Birds				Mammals			
Species	Temp (°C)	pH	P_{50}	Species	Temp (°C)	pH	P_{50}
Rhea[a]	41.0	7.5	30.5	Human[e]	37.0	7.4	28.0
Rhea[b]	41.0	7.5	20.7	Human[m]	37.0	7.5	24.0
Pheasant[b]	41.0	7.5	34.7	Human[c]	37.5	7.1	28.0
Pheasant[a]	41.0	7.5	42.2	Dog[m]	37.0	7.5	32.0
Chicken[c]	40.0	7.1	50.0	Dog[c]	37.5	7.1	32.0
Chicken[b]	41.0	7.5	35.1	Horse[m]	37.5	7.5	24.0
Chicken[d]	37.5	7.1	56.0	Mule[m]	37.5	7.5	24.0
Chicken[e]	37.0	7.4	25.4	Cow[m]	37.5	7.5	25.0
Pigeon[b]	41.0	7.5	29.5	Sheep[m]	37.5	7.5	27.0
Pigeon[d]	40.0	7.1	40.0	Goat[m]	37.5	7.5	29.0
Sparrow[b]	41.0	7.5	41.3	Pig[m]	37.5	7.5	34.0
Fulmar[f]	42.0	7.4	60.5	Cat[m]	37.5	7.5	40.0
Quail[b]	41.0	7.5	42.4	Opossum[m]	37.5	7.5	43.0
Mallard[b]	41.0	7.5	35.2	Rabbit[m]	37.5	7.5	30.0
Herring gull[f]	42.0	7.4	56.5	Hamster[m]	37.5	7.5	28.0
Pekin duck[g]	41.0	7.4	48.0	Deer mouse[m]	37.5	7.5	48.0
Muscovy duck[h]	41.0	7.5	41.7	Wood mouse[m]	37.5	7.5	49.0
Domestic duck[d]	37.5	7.1	45.0	Cotton rat[m]	37.5	7.5	39.0
Canada goose[g]	41.0	7.3	51.0	Wood rat[m]	37.5	7.5	33.0
House finch[i]	41.0	7.5	37.4	Guanaco[j,*]	37.0	7.4	22.5
Cassin's finch[d]	41.0	7.5	39.5				
Rosy finch[i,*]	41.0	7.5	31.0				
Bar-headed goose[g,*]	41.0	7.4	36.0				
Bar-headed goose[j,*]	37.0	7.4	29.7				
Bar-headed goose[k,*]			27.2				
Ruppell's griffon[l,*]			16.4				
Black vulture[l,*]			21.3				

[a] Boggs and Birchard (1982).
[b] Lutz *et al.* (1974).
[c] Morgan and Chichester (1935).
[d] Christensen and Dill (1935).
[e] Isaacks (1985).
[f] Clausen *et al.* (1971).
[g] Black and Tenney (1980b).
[h] Holle *et al.* (1977).
[i] Clemens (1990).
[j] Bauer and Petschow (1976).
[k] Black *et al.* (1978).
[l] Weber *et al.* (1988).
[m] Schmidt-Nielsen and Larimer (1958).
* high-altitude-adapted species.

TABLE X

ALLOMETRIC EQUATIONS AND THEIR VALUES FOR THE HALF-SATURATION PRESSURE (P_{50}, IN TORR) IN BIRDS AND MAMMALS OVER A RANGE OF BODY MASS (M_b, IN G)[a]

	Birds	Mammals	Bird/Mammal
Allometric equation	$P_{50} = 55.4\ M_b^{-0.079}$	$P_{50} = 50.3\ M_b^{-0.054}$	
Body mass (g)			
10	46.2	44.5	1.04
100	38.5	39.3	0.98
1,000	32.1	34.7	0.93
10,000	26.8	30.6	0.88
100,000	22.3	27.0	0.83

[a] Allometric equations from Lutz *et al.* (1974).

the lung. It is unknown whether further increases in ventilation would result in additional hyperventilation (decreased Pa_{CO_2}) and thus even higher arterial pH and lower P_{50} values.

Oxygen affinity of the blood increases with decreased body temperature (Maginniss, 1985; Pinshow *et al.*, 1985), which has been shown to decrease in resting hypoxic birds during cold exposure (Bernstein, 1989; Barnas and Rautenberg, 1990). It is not known if flight at high altitude results in decreased body temperature. However, Bar-headed geese flying over the Himalayan Mountains are exposed to extreme hypoxia and low temperatures. The inhalation of this cold air could decrease temperature of the blood at the gas–blood interface and thereby increase oxygen affinity of the blood.

Most low-altitude-adapted birds exposed to chronic hypoxia increase both their hemoglobin concentrations and packed cell volumes, as do low-altitude-adapted mammals (Weathers and Snyder, 1974; Black and Tenney, 1980a; Weinstein *et al.*, 1985). However, polycythemia does not occur in Bar-headed geese, even after a 4-wk exposure to a simulated altitude of 5640 m (Black *et al.*, 1978). Although polycythemia may be of value to the animal during acute exposures to hypoxia, prolonged polycythemia may place a burden on cardiac output by increasing the work load on the heart (Murry *et al.*, 1963). Therefore, the lack of polycythemia may be an adaptation for high-altitude flight.

d. Regional Blood Flow. Blood flow in birds, like that in mammals, is redistributed toward oxygen-dependent organs during exercise and exposure to hypoxia. The blood flow in swimming Tufted ducks increases to the working muscles and heart, whereas blood flow to the nonworking muscles (flight muscles), parts of the gastrointestinal tract, and pancreas is reduced (Butler *et al.*, 1988). Blood flow to the adrenals, spleen, respiratory muscles, liver, and brain was unaffected in these exercising birds (Butler *et al.*, 1988). In contrast, exposure of Bar-headed geese and Pekin ducks to hypoxia resulted not only in increased blood flow to the heart, but also to the brain, whereas blood flow to the skeletal muscles, kidneys, and gut of Pekin ducks was reduced (Faraci *et al.*, 1985). However, in the Bar-headed goose, with its high-oxygen-affinity hemoglobin, redistribution of blood flow during hypoxic exposure was less pronounced, with blood flow to the skeletal muscles being similar to that seen in normoxia (Faraci *et al.*, 1985).

Perhaps the most important factor in the greater tolerance of hypoxia in birds, compared to mammals, is the regulation of cerebral blood flow. In mammals, decreasing Pa_{O_2} results in hyperventilation that, in turn, leads to decreased Pa_{CO_2}. The decrease in Pa_{CO_2} reduces cerebral blood flow, thus attenuating further increases in cerebral blood flow due to hypoxic stimulus (Grubb *et al.*, 1977, 1979; Faraci, 1988). Severe hypocapnia in cats can completely abolish cerebral blood flow (Davies *et al.*, 1986), and decreased cerebral perfusion appears to affect both cerebral sensory and motor functions (West, 1984).

Reduction of Pa_{O_2} increases cerebral blood flow in birds, just as it does in mammals. However, in contrast to mammals, Bar-headed geese can maintain, and even increase, cerebral blood flow in the presence of severe hypocapnia (Faraci and Fedde, 1986). This response enhances the hypoxic tolerance of all birds, but may be particularly invaluable to high-altitude-adapted species. During hypoxia, high-altitude species that possess hemoglobin with a high oxygen affinity can maintain, or even increase, oxygen delivery to the brain with little or no increase in cerebral blood flow (Faraci *et al.*, 1984a). Species that do not fly at high altitudes, such as the Pekin duck, have lower arterial oxygen concentrations when exposed to hypoxia, but can still maintain oxygen delivery to the brain; they do so by dramatic increases in cerebral blood flow (Grubb *et al.*, 1979; Faraci and Fedde, 1986). [N.B.: Although Pekin ducks are not strong fliers, they are conspecifics of Mallard ducks (*Anas platyrhynchos*) that migrate at altitudes in excess of 6400 m.]

e. Rheology. The nonnucleated red blood cells of mammals are well adapted for a high oxygen carrying capacity without creating high viscous resistance. However, the nucleated red blood cells of birds apparently lack this advantage (Gaehtgens, 1990). Nucleated red blood cells are less deformable than nonnucleated red blood cells, have a lower stability of orientation during passage through the capillaries, and occupy a larger fraction of the cross section of the capillary (Gaehtgens *et al.*, 1981a). All of these factors increase the effects of hemodynamic disturbance and thus have profound effects on capillary flow hemodynamics (Gaehtgens *et al.*, 1981a). A further disadvantage for birds is the high relative viscosity of avian blood. Red cell suspensions of Pekin ducks have a relative viscosity four to six times higher than that of human cell suspensions, when tube diameter and hematocrits entering the tube are kept equal (Gaehtgens *et al.*, 1981b). If the results of these *in vitro* experiments are applicable to the microcirculation in living ducks, a significantly higher rheological component of flow resistance, compared to that of humans, may exist in birds (Gaehtgens *et al.*, 1981b).

Birds possess several adaptations that compensate for their apparently higher flow resistances. The skeletal muscles, cardiac muscle, and brain tissues of birds have greater capillary densities than do the same tissues in mammals (Faraci, 1991). This increase in capillary number would act to increase the conductivity of the capillary beds, thus compensating for higher flow resistance, and a decrease in capillary hematocrit would aid in offsetting high flow resistance. Capillary hematocrit has been measured in only one avian species, the Pekin duck, and was nearly half that measured in various mammals (Gaehtgens *et al.*, 1981b). Although the rheological properties of avian blood appear to be a disadvantage for oxygen transport, compared to those of mammalian blood, birds have apparently been able to counterbalance this drawback by adaptations in microvascular architecture.

3. *Tissue Specializations*

a. Fiber Type Composition. Wing beat frequencies are high during flight for most birds: hummingbirds, 15–80 Hz; pigeons, 6 Hz; barnacle geese, 5 Hz; largest flying birds, 2–3 Hz (Greenewalt, 1962; Torre-Bueno and Larochelle, 1978; Butler and Woakes, 1980; Marsh, 1984). As a result, the *pectoralis* and *supracoracoideus* muscles (the flight muscles) of most birds consist only of fast-twitch muscle fibers. These fibers must be able to maintain high wing beat frequencies without fatiguing; thus, the flight muscles of birds contain primarily fast-twitch oxidative (Type IIA) fibers. These fast-twitch fibers contain large volumes of mitochondria, are surrounded by numerous capillaries, and are capable of utilizing fats as fuel.

In passerine birds, the *pectoralis* muscle consists almost entirely of Type IIA fibers (Rosser and George, 1986). In heavier birds, both Type IIA and fast-twitch glycolytic (Type IIB) fibers exist. The Type IIB fibers are concentrated primarily in the superficial region of the *pectoralis major* muscle, with comparatively few fibers of this type occurring in the deeper region of the muscle (George and Berger, 1966; Kaiser and George, 1973; Rosser and George, 1986). These fibers are adapted for brief bursts of power (George and Berger, 1966) and are utilized during intense activities, such as takeoff (Parker and George, 1973, 1976).

The cross-sectional areas of the Type IIA fibers in avian flight muscle, even in large birds, are smaller than those found in the locomotor muscles of most mammals (Mathieu-Costello, 1990). This smaller size increases the capillary surface area to fiber surface area and thus decreases the diffusion distance for oxygen and substrates. Further, the small cross-sectional area of each fiber decreases the diffusion distance from the periphery of the fiber to mitochondria located near its center. Such adaptations could enhance oxygen transfer to the muscle mitochondria.

b. Capillary Volume. Adaptations exist in birds that enhance the transport of oxygen from capillary blood to cellular mitochondria. Among these is high capillary density in the skeletal muscle, cardiac muscle, and brain tissues (Lierse, 1963; Stoeckenius, 1963; Folkow *et al.*, 1966; Rakusan *et al.*, 1971; Mathieu-Costello, 1990). The flight muscles of pigeons and hummingbirds contain large numbers of capillary branches running perpendicular to the fiber axis (Mathieu-Costello, 1990). This branching results in an increased capillary-to-fiber perimeter ratio, an indicator of the capillary surface available for oxygen and metabolite transfer per unit area of fiber surface (Mathieu-Costello, 1990). This ratio in the *pectoralis* muscle of the pigeon and hummingbird is greater than that found in mammalian limb muscles and in rat heart (Mathieu-Costello, 1990). Differences in capillary density of the *pectoralis* muscles also occur among avian species. Passerine birds that migrate over long distances typically have greater capillary densities, smaller fiber diameters, and thus decreased diffusion distances in the flight muscles than do nonmigratory birds (Table XI). Further increases in the capillary-to-fiber perimeter ratio may be induced by exercise and/or hypoxia, as capillary growth in skeletal and cardiac muscle may be triggered by these stimuli (Andersen and Henricksson, 1977; Klausen *et al.*, 1981; Hudlicka *et al.*, 1984; Snyder *et al.*, 1984; Burton and Barclay, 1986; Vetterlein *et al.*, 1989).

c. Myoglobin. Myoglobin, an oxygen binding protein in muscle fibers, also may play a vital role in the movement of oxygen from capillaries to muscle mitochondria (Gayeski and Honig, 1986; Wittenberg and Wittenberg, 1989). The function of myoglobin may become in-

TABLE XI

CAPILLARY DENSITY (CD), MEAN FIBER AREA (FA), AND MEAN DIFFUSION LENGTH (DL) IN THE *PECTORALIS* MUSCLE OF LONG-DISTANCE, SHORT-DISTANCE, AND PARTIALLY MIGRATORY BIRDS[a]

Species	CD (mm^2)	FA[b] (μm^2)	DL (μm)
Long-Distance Migrants			
Sand martin	2003	900	10.4
Sedge warbler	1791	950	10.8
Reed warbler	2020[c]	700	9.8[c]
Willow warbler	1959[c]	750	9.7[c]
Pied flycatcher	1644	700	11.1
Spotted flycatcher	1704	800	10.9
Red backed shrike	1700	900	10.6
Mean value	1832	814	10.5
Short-Distance Migrants			
Robin	1607	1100	11.5
Blackbird	1215	1200	13.9
Goldcrest	2204	700	9.4
Reed bunting	1535	1300	11.8
Mean value	1640	1075	11.8
Partial or Nonmigrants			
Marsh tit	1748[c]	1100[c]	11.1[c]
Willow tit	1662	1000	11.0
Great tit	1497	1000	11.8
Yellowhammer	1416	1500	12.5
Mean value	1581	1150	11.6

[a] From Lundgren and Kiessling (1988).
[b] Approximate value for fast-twitch oxidative fibers.
[c] Mean of spring and fall migratory values.

creasingly important under conditions of lowered arterial oxygen tensions, like those encountered during strenuous exercise or at high altitude (Covell and Jacquez, 1987; King *et al.*, 1987).

The concentration of myoglobin is higher in the *pectoralis* muscle of low-altitude-adapted, strong-flying birds compared to weak-flying, low-altitude-adapted birds. Myoglobin concentrations appear to be even higher in the *pectoralis* muscles of high-altitude-adapted, strong-flying birds (Table XII). Myoglobin has a very high affinity for oxygen (P_{50} =

TABLE XII

MYOGLOBIN CONCENTRATIONS IN THE *PECTORALIS* AND CARDIAC MUSCLES OF VARIOUS AVIAN SPECIES

Species[a]	Myoglobin concentration ($mg\ g^{-1}$ wet weight) Heart	*Pectoralis*
	Low-Altitude, Weak-Flying Birds	
Domestic chicken	2.8[b]	0.3[c]
	3.6[d]	2.0[d]
	1.3[e]	1.0[e]
Farm-raised quail	6.8[e]	0.5[e]
Wild quail	3.1[e]	2.8[e]
Ruffed grouse	—	0.2[f]
Spruce grouse	—	2.5[f]
Pheasant	—	0.5[g]
	Low-Altitude, Strong-Flying Birds	
Pigeon	3.0[e]	4.7[e]
	3.5[h]	2.2[i]
	—	7.2[g]
European starling	2.0[e]	2.0[e]
Starling (R)	2.1[j]	2.1[j]
Starling (A)	7.2[j]	9.5[j]
Seagull	6.4[i]	5.5[i]
Hummingbird[k]	2.8[l]	3.4[l]
Tufted duck (UT)	7.0[m]	4.7[m]
Tufted duck (NT)	6.8[m]	7.4[m]
	High-Altitude, Strong-Flying Birds	
Grey heron	—	6.3[n]
Ural owl	—	5.9[n]
Andean cock-of-the rock	—	8.5[n]
Bar-headed goose	6.0[o]	6.4[o]
	Diving Birds	
Adelie penguin	—	29.2[p]
	—	36.0[q]
Gentoo penguin	—	46.2[p]
Chinstrap penguin	—	32.0[p]
Rock hopper penguin	—	14.7[n]
King penguin	—	35.9[n]
Little penguin	11.0[q]	28.0[q]

(*continued*)

TABLE XII (*Continued*)

	Myoglobin concentration (mg g^{-1} wet weight)	
Species[a]	Heart	*Pectoralis*
Emperor penguin	—	24.0[q]
Pigeon guillemot	16.1[r]	21.6[r]
Common murre	—	14.1[g]
Atlantic puffin	—	12.5[g]

[a] R, values from Starlings resting on wintering grounds; A, values from Starlings that had just arrived from migration; UT, values from untrained Tufted ducks; and NT, values from trained Tufted ducks.
[b] Nishida (1976).
[c] Akeson *et al.* (1960).
[d] Nishida and Nishida (1978).
[e] Pages and Planas (1983).
[f] Thomas (1985).
[g] Davis and Guderly (1987).
[h] Schuder *et al.* (1979).
[i] Lawrie (1953).
[j] Palacios *et al.* (1984).
[k] Average value from 11 different species of hummingbirds.
[l] Johansen *et al.* (1987).
[m] Stephenson *et al.* (1989).
[n] Enoki *et al.* (1987).
[o] Saunders and Fedde (1991).
[p] Weber *et al.* (1974).
[q] Mill and Baldwin (1983).
[r] Haggblom *et al.* (1988).

2.5 Torr; Johansen *et al.*, 1987) and thus may aid the release of oxygen from the high affinity hemoglobin in blood by maintaining a steep oxygen partial pressure gradient between capillaries and muscle mitochondria. The steep gradient is apparently maintained as a result of the immediate binding of oxygen molecules by myoglobin as they pass through the muscle membrane, thereby maintaining a low P_{O_2} within the muscle cells. Further, myoglobin may actually mediate the delivery of oxygen from the muscle membrane to muscle mitochondria (Wittenberg and Wittenberg, 1987). The role of myoglobin in the muscle tissue may significantly contribute to the muscles' ability to maintain maximal work output under conditions of hypoxia and exercise.

Most of the myoglobin measurements reported in the literature have been performed on birds that were not adapted for exercise or hypoxia. Increased myoglobin concentrations in skeletal and cardiac muscles of mammals have been associated with prolonged endurance exercise (Pattengale and Holloszy, 1967; Hagler *et al.*, 1980; Hickson, 1981; Butler and Turner, 1988; Saunders and Fedde, 1991), as well as prolonged exposure to high altitude (Hurtado *et al.*, 1937; Turek *et al.*, 1973; Reynafarje *et al.*, 1975). Thus, myoglobin concentrations in the *pectoralis* and cardiac muscles of Bar-headed geese in their native environment may be higher than those reported (Saunders and Fedde, 1991).

d. Mitochondrial Volume. The aerobic performance of muscle is intrinsically linked to its mitochondrial content. Some of the highest values for mitochondrial volume density in skeletal muscle have been found in the flight muscles of birds. In the *pectoralis* muscle of hummingbirds, mitochondria occupy nearly 35% of the total fiber volume (Suarez *et al.*, 1991). Further, volume-specific respiration rates of mitochondria from flight muscles of hummingbirds were found to be about two times greater than those from mammalian locomotor muscle (Suarez *et al.*, 1991). The only morphometric difference between the mitochondria from these two sources was an increased (nearly twofold) mitochondrial membrane surface area per unit mitochondrial volume in hummingbirds (Suarez *et al.*, 1991).

It has been suggested that the higher volume-specific respiration rates of hummingbird mitochondria *in vivo*, compared to those of mammals, result from higher capacities of oxygen and substrate delivery, as no differences were found in the *in vitro* respiration rates of the inner mitochondrial membranes of hummingbirds and mammals. The increased capacity for oxygen delivery at the tissue level in hummingbird muscle may occur because of clustering of the mitochondria near the sarcolemma, adjacent to capillaries. Such clustering also has been shown in the flight muscle of Bar-headed and domestic geese (Swatland, 1984; Fedde *et al.*, 1985).

e. Enzymatic Properties. The ability of muscle fibers to produce the large quantities of adenosine triphosphate (ATP) needed for flight is dependent not only on their mitochondrial volume, but also on the activities of the oxidative enzymes within the mitochondria. High activities of the enzymes within the citric acid cycle are indicative of the endurance capacity of the muscle. The activities of many oxidative enzymes in the flight muscles of birds are among the highest found in vertebrate muscle.

Citrate synthase catalyzes near-equilibrium reactions in the citric acid cycle of vertebrate muscle (Alp *et al.*, 1976) and thus is often used to measure the aerobic potential of the muscles. The highest citrate synthase activity in vertebrate muscle is found in the flight muscle of the Rufous hummingbird (Suarez *et al.*, 1986, 1990) (Table XIII). High citrate synthase activities have also been reported for the *pectoralis* muscles of several other avian species, but the activity of this enzyme is low in the domestic Chicken (Table XIII). The values for citrate synthase activity in the *pectoralis* muscles of some birds are higher than those found in cardiac muscle (Table XIII).

B. Thermoregulation and Water Conservation

Body temperature in birds increases dramatically during flight, with core temperature increasing by as much as 2–4°C (Hart and Roy, 1967; Torre-Bueno, 1976; Hudson and Bernstein, 1981; Hirth *et al.*, 1987). It has been suggested that heat retention during flight is an adaptation that enhances muscular power output and increases the unloading of oxygen to working muscle (Torre-Bueno, 1976). However, this extreme hyperthermia may be only slightly below the body temperature that would be lethal to the bird and, therefore, must be closely regulated.

The predominant mechanism for heat dissipation in flying birds is convection from airflow over the body surface. In flying Herring gulls, about 80% of heat produced is dissipated through the webbed feet, in which blood flow increases by about 3.5 times above that during rest (Baudinette *et al.*, 1976). Even birds that lack webbed feet (e.g., pigeon) utilize their feet for heat dissipation (Biesel and Nachtigall, 1987).

Heat loss via respiratory evaporation at an ambient temperature of about 20°C is less than 25% of the total heat loss for most birds (Dawson, 1982). However, respiratory evaporation results in a loss of water that is greater than metabolic water production, even at moderate ambient temperatures (Tucker, 1968b; Biesel and Nachtigall, 1987). This imbalance produces a water deficit that could have a profound influence on long-distance, nonstop migrants. Tucker (1968b) calculated metabolic water production and determined the water loss of Budgerigars during level flight at 35 km h^{-1} in a wind tunnel. He assumed the fuel of these flying birds was 75% fat and 25% carbohydrate, resulting in the formation of 13 mg H_2O g^{-1} body wt h^{-1}. Water was lost via evaporation at a rate of 20 mg H_2O g^{-1} body wt h^{-1}, leading to a negative body water balance (excluding that from excretion) of 7 mg H_2O g^{-1} body wt h^{-1}, or a net loss of 0.7% of the body weight per hour. It has been shown that birds can easily sustain a water loss of 15% of total body weight (Bartholomew and Cade, 1956).

TABLE XIII

ACTIVITIES OF CITRATE SYNTHASE IN VERTEBRATE MUSCLE

Animal	Muscle[a]	Citrate synthase activity (μmol min^{-1} g^{-1})
Pisces		
Dogfish[b]	Red	35.0
(*Scylliorhinus canicula*)	White	1.7
	Heart	46.0
Trout[b]	Red	50.0
(*Salmo gairdneri*)	White	5.2
	Heart	38.0
Amphibia		
Frog[b]	*Sartorius*	10.0
(*Rana temporaria*)	Heart	40.0
Toad[b]	*Sartorius*	7.0
(*Xenopus laevis*)	*Gastrocnemius*	12.0
Aves		
Chicken[b]	*Pectoralis*	6.8
(*Gallus gallus*)	Heart	51.0
Pigeon[b]	*Pectoralis*	115.0
(*Columba livia*)	Heart	127.0
Pigeon[c]		
(*Columba livia*)	*Pectoralis*	51.3
	Supracoracoideus	29.1
	Sartorius	7.4
Tufted duck[d]	*Pectoralis* (UT)	86.5
(*Aythya fuligula*)	(NT)	150.8
	Heart (UT)	108.0
	Heart (NT)	128.1
	Gastrocnemius (UT)	30.8
	Gastrocnemius (NT)	94.8
	Semitendinosus (UT)	23.8
	Semitendinosus (NT)	65.5
Sparrow[b]	*Pectoralis*	112.0
(*Passer domesticus*)	Heart	120.0
Robin[c]	(M)	200.0
(*Erithacus rubecula*)	(B)	120.0
Blackbird[e]	(M)	140.0
(*Turdus mercula*)	(B)	100.0
Sedge warbler[e]	*Pectoralis* (M)	210.0
(*Acrocephalus schoenobaenus*)	(B)	190.0
Reed warbler[e]	*Pectoralis* (M)	215.0
(*Acrocephalus scirpaceus*)	*Pectoralis* (B)	160.0
Great tit[e]	*Pectoralis* (M)	140.0
(*Parus major*)	(B)	130.0

(continued)

TABLE XIII (*Continued*)

Animal	Muscle[a]	Citrate synthase activity (μmol min^{-1} g^{-1})
Reed bunting[e]	*Pectoralis* (M)	170.0
(*Emberiza schoeniclus*)	(B)	150.0
American goldfinch[f]	*Pectoralis* (summer)	210.0
(*Carduelis tristis*)	*Pectoralis* (winter)	203.0
	Mixed leg (summer)	88.2
	Mixed leg (winter)	71.1
Gray catbird[g]	*Pectoralis*	201.0
(*Dumetella carolinensis*)	*Supracoracoideus*	90.3
Rufous hummingbird[h]	*Pectoralis* and	343.3
(*Selasphorus rufus*)	*supracoracoideus*	190.3
	Heart	
Pheasant[b]	*Pectoralis*	14.4
(*Phasianus colchicas*)	Heart	99.0
Pheasant[c]	*Pectoralis*	10.5
(*Phasianus cochicas*)	*Supracoracoideus*	9.7
	Sartorius	5.9
Atlantic puffin[c]	*Pectoralis*	46.3
(*Fratercula arctica*)	*Supracoracoideus*	31.6
	Sartorius	2.2
Common murre[c]	*Pectoralis*	51.1
(*Uria aalge*)	*Supracoracoideus*	27.0
	Sartorius	1.5
Little penguin[i]	*Pectoralis*	20.8
(*Eudyptula minor*)	*Supracoracoideus*	21.5
Mammalia		
Lab mouse[b]	Heart	146.0
Lab rat[b]	Heart	96.0
Rabbit[b]	Heart	69.0
Sheep[b]	Heart	58.0
Pig[b]	Heart	50.0
Ox[b]	Heart	54.0

[a] UT, untrained Tufted ducks; NT, naturally trained Tufted ducks; M, migratory; and B, breeding.
[b] Alp *et al.* (1976).
[c] Davis and Guderley (1987).
[d] Butler and Turner (1988).
[e] Lundgren and Kiessling (1985).
[f] Marsh and Dawson (1982).
[g] Marsh (1981).
[h] Suarez *et al.* (1986).
[i] Mill and Baldwin (1983).

Therefore, if water loss is assumed to occur only by way of respiratory evaporation, then the Budgerigar could fly for approximately 21 h before 15% dehydration would occur.

It has been estimated that some migratory birds may fly for 50 h or more without stopping for water (Moreau, 1961; Nisbet *et al.*, 1963). These birds must possess either physiological or behavioral adaptations that allow them to lose water more slowly than is the case with the Budgerigar. Further, long-distance migrants may be better able to withstand a greater total water loss than the Budgerigar. Very little is known about possible physiological adaptations for water conservation in long-distance migrants.

The altitude at which birds fly may influence respiratory water loss. It has been suggested that birds fly at altitudes that optimize the balance between the gain of metabolic water and the loss of water via evaporation (Torre-Bueno, 1978b). To this extent, higher altitudes may be beneficial, as respiratory water loss decreases with lower environmental temperatures that occur at higher altitudes.

It has been shown experimentally that at lower temperatures the production of metabolic water can actually equal the rate of respiratory water loss in Black ducks and Starlings during flight (Berger *et al.*, 1971; Torre-Bueno, 1978b). Decreased evaporative water loss at lower ambient temperatures is apparently due to two different mechanisms. The first is a countercurrent exchange mechanism. As cold air is inspired and passes through the nasal passages, it is warmed to body temperature and saturated with water vapor. The epithelium of the nasal passages cools as it loses heat to the inspired air. In the European coot at ambient temperatures between 10° and 35°C, the difference between ambient air temperature and expired gas temperature is 5°C or less (Brent *et al.*, 1984). If the ambient temperature decreases to −10°C or less, the temperature of the nasal epithelium near the nasal openings can be nearly 0°C (Brent *et al.*, 1984). As the warmed air is then exhaled over the cooler nasal surfaces, water condenses onto the nasal mucosa. The condensed water formed at the end of expiration is then used to moisten the inspired air, thereby conserving body water. The importance of this mechanism for water and heat conservation has been shown in a variety of birds (Schmidt-Nielsen *et al.*, 1970).

The second mechanism for conserving body water involves increasing pulmonary oxygen extraction, discussed previously, with exposure to cold temperatures (Bucher, 1981; Brent *et al.*, 1983, 1984; Bech *et al.*, 1984). Increased oxygen extraction decreases the ventilatory requirements needed to maintain oxygen consumption. This, in turn, would decrease the loss of water due to evaporation.

IV. Physiological Adaptations for Diving

Many aquatic avian species dive to obtain food, most notably penguins and diving ducks. There also is one diving passerine bird, the Dipper (*Cinclus mexicanus*), which is the smallest diving bird, weighing about 50 g.

Penguins dive to the greatest depth of any birds, with the Emperor penguin holding the record for the deepest known dive at 265 m (Kooyman *et al.*, 1971). Although these birds are capable of diving to great depths, most dives of Emperor penguins are to much shallower depths (Piatt and Nettleship, 1985; Kooyman and Ponganis, 1990).

The duration of most dives is short (Butler and Jones, 1982). Gentoo and Chinstrap penguins usually remain submerged between 1.5 and 2.5 min (Trivelpiece *et al.*, 1986), but longer dives (2–8 min) have been recorded for Emperor penguins (Kooyman and Ponganis, 1990). These dives are longer than those recorded for female fur seals and sea lions (2–4 min) (Gentry and Kooyman, 1986).

Most, if not all, of the energy formation (ATP) during the dives of penguins and most other diving birds results from aerobic metabolism (Kooyman *et al.*, 1973; Millard *et al.*, 1973; Butler and Woakes, 1982). After a dive, the Emperor penguin surfaces for only 2 or 3 min before diving again (Kooyman and Ponganis, 1990). Gentoo and Adelie penguins may spend less than 2 s at the surface between dives, and maintain this behavior for several hours (Millard *et al.*, 1973). Such short interdive durations would not allow for full recovery from anaerobic metabolism prior to the next dive, supporting the hypothesis that the dives are fully aerobic.

A. Pulmonary Specializations

The ability of penguins to undertake long aerobic dives probably is related to the size of their mass-specific oxygen stores (Kooyman and Ponganis, 1990). In penguins, oxygen is stored in the air sacs (29% of the total), in the blood (38% of the total), and in the muscle (33% of the total) (Kooyman and Ponganis, 1990). The total oxygen store of the Emperor penguin has been calculated to be 58 ml O_2 kg^{-1}, which is larger than that calculated for the Weddell seal (50 ml O_2 kg^{-1}; Kooyman, 1981), an animal capable of long breath holds. This greater oxygen store in the penguins derives primarily from the greater oxygen storage capacity of the lung–air sac system of these birds, as the birds inspire before diving, whereas the mammals expire.

Not only does the lung–air sac system of birds store a large volume of oxygen, but it also allows gas exchange to occur during the dive. Most diving mammals exhale prior to diving, thereby removing nitrogen from the lungs and preventing desolubilized nitrogen bubble formation in the blood (the bends) after a dive. However, penguins, as well as diving ducks, hyperventilate before diving, possibly to increase their oxygen stores (Kooyman *et al.*, 1971; Butler and Woakes, 1979), inspire before diving, and do not exhale under water (Kooyman *et al.*, 1971). Evidence suggests that gas exchange occurs during dives in penguins. In Adelie and Gentoo penguins compressed in a water-filled chamber to pressures equivalent to depths of 30 and 68 m, oxygen concentrations of the thoracic air sac fell during compression, indicating that gas was exchanged between the air sacs, lungs, and blood during compression (Kooyman *et al.*, 1973). Furthermore, in these same two species, no discernable difference was detected in arterial P_{O_2} before and after diving, suggesting gas exchange between the lungs and blood throughout the dive (Millard *et al.*, 1973).

Gas exchange between the lung and blood may be aided by the effects of compression during diving. For every 10 m of depth, pressure increases 1 atm (760 Torr). For a bird diving to 40 m, the pressure of the gas in its air sacs would be about 5 atm, or 3800 Torr. The partial pressure of oxygen in the air sacs (assuming 21% oxygen) would then be about 800 Torr. This high driving pressure should enhance the movement of oxygen from the lung into the blood. Even if the concentration of oxygen dropped to just 2% in the air sac, the driving pressure for oxygen would still be 76 Torr.

A significant deleterious effect of compression results from the pressurization of nitrogen in the gas within the lung. With higher pressures, more nitrogen can become solubilized in the blood and body fluids. During surfacing from deep dives, there is a danger of nitrogen bubble formation in the blood vessels, especially those in the brain. Nitrogen bubble formation in the blood leads to the bends. For most diving birds, the depths (and thus, the pressures) reached, as well as the exposure times at those depths, are sufficiently low to minimize the probability of high nitrogen solubilization occurring in the blood (Kooyman and Ponganis, 1990). However, this is not the case for the King and Emperor penguins, which dive to depths greater than 200 m. Further, the dives of these birds are repetitive, a factor that has been shown to increase the risk of nitrogen bubble formation in human divers (Paulev, 1965).

A conflict exists between the need to prevent nitrogen bubble formation and to maintain adequate arterial oxygen tensions during deep dives. Nitrogen bubble formation in cats occurred upon decompression when the animals had blood nitrogen tensions of about 2250 Torr (Harvey *et al.*, 1944). Such high nitrogen pressures in the blood could occur in penguins after prolonged exposure to depths of about 30 m. Penguins are presumably no less susceptible to bubble formation than are other vertebrates. Thus, the partial pressure of nitrogen in the blood must be kept lower than that which produces bubbles during deep dives. To accomplish this, blood must be shunted away from the lungs to prevent gas exchange. It has been estimated that during dives greater than 100 m, the amount of shunt that must occur to prevent nitrogen tensions from exceeding 2250 Torr is about 98%, only allowing 2% of the cardiac output to exchange gases with the lungs, and resulting in a precarious fall in arterial oxygen tensions (Kooyman and Ponganis, 1990). The mechanism(s) by which diving penguins are able to maintain aerobic dives and low blood nitrogen partial pressures remain to be elucidated.

B. Cardiovascular Specializations

Many of the earliest diving studies were performed on animals that were forcibly submerged. Forced submersion resulted in decreased metabolism (Scholander and Irving, 1941; Anderson *et al.*, 1965; Murphy *et al.*, 1980), as well as bradycardia, immediately upon or shortly after submersion (Cohn *et al.*, 1968; Murrish, 1970; Kooyman *et al.*, 1973; Millard *et al.*, 1973; Blix and Berg, 1974; Butler and Woakes, 1982). These reactions to submersion, termed the diving reflex (Blix and Folkow, 1983), act to conserve oxygen by decreasing perfusion of the gut, kidneys, liver, and skeletal muscle (Johansen, 1964; Butler and Jones, 1971; Jones *et al.*, 1979), and to maintain oxygen delivery to the brain and heart (Johansen, 1964; Butler and Jones, 1971; Jones *et al.*, 1979; Bryan and Jones, 1980).

The diving reflex does not occur in all birds during natural dives. In Jackass, Adelie, and Gentoo penguins, heart rate does not decrease lower than that at rest during dives of 60 s or less (Butler and Woakes, 1982). The same has been found for Humboldt penguins during dives of 30 s (Butler and Woakes, 1984), and Emperor penguins diving longer than 10 min (Kooyman and Ponganis, 1990). Furthermore, bradycardia is not maintained during natural dives of the Double-crested cormorant (*Phalacrocorax auritis*) (Kanwisher *et al.*, 1981), Tufted duck, or Pochard (*Aytha ferina*) (Butler and Woakes, 1976, 1979). However, bradycardia has been shown during the natural dives of the Gentoo and Adelie penguins (Millard *et al.*, 1973) and the Dipper (Murrish, 1970).

In many diving birds (Gentoo and Adelie penguins, as well as the Dipper), increases in heart rate and cardiac output occur just prior to surfacing (Murrish, 1970; Millard *et al.*, 1973). This tachycardia is often accompanied by markedly increased ventilation at the surface (Millard *et al.*, 1973). Because of the short duration of the interdive period, such a response may be beneficial in increasing the rate at which oxygen stores can be replaced prior to the next dive.

There appears to be little adaptation in the respiratory properties of the blood of diving birds compared to in terrestrial avian species (Lenfant *et al.*, 1969). Although there are few comparative data, it does seem that penguins might have slightly higher oxygen-carrying capacities and blood volumes than do terrestrial birds (Kooyman, 1975).

C. Tissue Specializations

The most striking tissue specialization of diving birds, particularly the penguins, is the high concentration of myoglobin in the skeletal and cardiac muscle (Table X). However, these high myoglobin concentrations are not greater than those in the skeletal and cardiac muscles of diving mammals. Furthermore, myoglobin affinity for oxygen does not differ between diving penguins and other vertebrates (Weber *et al.*, 1974). High myoglobin concentrations in both diving birds and mammals are thought to serve as oxygen stores (Scholander, 1940; Weber *et al.*, 1974; Baldwin *et al.*, 1984; Davis and Guderley, 1987).

Aside from the high concentrations of myoglobin in the skeletal and cardiac muscles, there are very few other tissue specializations in diving birds. Davis and Guderley (1990) stated that "diving does not require major modifications of the overall metabolic design of the muscle used by a bird capable of sustained aerobic flight." This observation would seem to be supported by the finding that activities of several glycolytic enzymes, including lactate dehydrogenase, phosphorylase, and phosphofructokinase in the *pectoralis* muscles of the Little penguin (*Eudyptula minor*), Common murre (*Uria aalge*), and Atlantic puffin (*Fratercula arctica*), are not appreciatively different from those in the *pectoralis* muscles of the pigeon or Mallard duck (Mill and Baldwin, 1983; Davis and Guderley, 1990). In contrast, the activity of the oxidative enzyme citrate synthase was found to be lower in diving birds than in flying birds (Mill and Baldwin, 1983; Davis and Guderly, 1990) (Table XIII).

Some apparent differences do occur among diving birds in anaerobic enzyme activities within the *pectoralis* muscles. Mill and Baldwin (1983) investigated the distributions of lactate dehydrogenase isoenzymes, myoglobin content, buffering capacity, and fiber diameter in the *pectoralis* and *supracoracoideus* muscles of seven species of penguins. They ranked the anaerobic capacities of the *pectoralis* and *supracoracoideus* muscles as being highest in the King penguin, followed by the Emperor, Adelie, Gentoo, Royal and Rockhopper, and Little penguins. This order tends to mirror the diving capabilities for length and depth of these birds (Baldwin *et al.*, 1984).

V. Conditioning and Deconditioning in Birds

Both conditioning and deconditioning occur in the muscles of birds. Prolonged physical conditioning of captive birds increases myoglobin concentrations and oxidative enzyme activities within the exercised muscles (Butler and Turner, 1988; Brackenbury and Williamson, 1989; Saunders and Fedde, 1991). On the other hand, confinement of captive birds decreases myoglobin concentrations in the heart and *pectoralis* muscles of pigeons (Catlett *et al.*, 1978; Pages and Planas, 1983) and decreases myoglobin concentrations and citrate synthase activities in the heart and *pectoralis* muscles of Tufted ducks (Butler and Turner, 1988). Thus, muscle usage appears to influence certain biochemical factors within avian skeletal and cardiac muscles.

Migration is a strenuous and demanding form of exercise for the flight muscles of birds. However, migrations usually occur only twice a year and require relatively little time to complete. In between the biannual migrations, there are prolonged periods of decreased flight muscle usage, and, in some cases (e.g., feather molt and incubation), natural inactivity of the flight muscle. It is likely that the flight muscles of birds decondition during these periods of decreased muscle use. If deconditioning occurs, is reconditioning of the flight muscles of these birds required prior to migration, or are the muscles conditioned during the actual migratory trip?

A. Seasonal Changes in Muscle Mass

Nearly all migratory birds show seasonal changes in muscle mass. Prior to long migrations, small passerine birds may double their body weight (Odum *et al.*, 1961; Fry *et al.*, 1970; Marsh, 1984), with a resulting increase in the power requirement for flight (Pennyquick, 1975). To compensate for this increased power requirement, the flight muscle mass of these birds enlarges (Fry *et al.*, 1970, 1972; Ward and Jones, 1977; Marsh, 1984). Increases in muscle mass associated with increases in body weight also occur in larger birds, such as the Cooper's hawk (Marsh and Storer, 1981), resulting from hypertrophy of existing muscle fibers (Marsh, 1984). Such adaptations suggest that the flight muscle mass of a bird may be regulated by the load placed upon it during flight (Marsh, 1984). Therefore, birds may, in effect, "power train" their flight muscles in a manner very similar to that of the limb muscles of human weight lifters.

Many migratory passerine birds undergo a twice-yearly cycle of nocturnal restlessness prior to migration known as *Zugunruhe*. Interestingly, *Zugunruhe* is associated with the premigratory fattening that occurs in migratory birds (King and Farner, 1963). During this time, the birds exhibit increased fluttering and hopping behavior during the night. Thus, not only is there an increased load on the flight muscle due to increased body weight, but also an increased use of the flight muscles. Therefore, *Zugunruhe* may further stimulate premigratory conditioning of the flight muscles.

In stark contrast to the increase in muscle mass of premigratory passerines, muscle mass of molting waterfowl (e.g., ducks, geese, swans) decreases dramatically. These birds may travel thousands of kilometers to reach their summer breeding grounds. After the breeding season, they quickly lose all of their flight feathers and thus the ability to fly (Bellrose, 1976). At this time, these birds may lose between 30 and 40% of their migratory flight muscle mass (Ankney, 1979, 1984; Raveling, 1979; Young and Boag, 1982; Bailey, 1985; DuBowy, 1985; Rosser and George, 1987; Gaunt *et al.*, 1990; Saunders and Klemm, 1991). Interestingly, the loss of flight muscle mass is usually associated with an increase in the mass of the leg muscles (Raveling, 1979; Young and Boag, 1982; Ankney, 1984; Bailey, 1985; DuBowy, 1985; Rosser and George, 1987; Gaunt *et al.*, 1990; Saunders and Klemm, 1991). However, like other waterfowl, the Eared grebe decreases the mass of the breast muscle, but has no concomitant increase in leg muscle mass (Gaunt *et al.*, 1990).

The seasonal change in flight and leg muscles may result from a use/disuse phenomenon (Young and Boag, 1982; DuBowy, 1985; George *et al.*, 1987; Rosser and George, 1987). However, attempts to correlate behavior with changes in muscle mass have failed to show either increased muscle use with increased muscle mass or decreased muscle mass with decreased usage (Ankney, 1984; Bailey, 1985; Gaunt *et al.*, 1990; Saunders and Klemm, 1991). In fact, it appears that the muscles of these birds may be changing in an anticipatory response to the upcoming demands to be placed on the muscles. An indication of such a response may be the appearance of apparently mitotically active satellite cells in the *pectoralis* muscles of Canada geese prior to the fall migration. The presence of these cells indicates a regenerative and repair process in the flight muscles, possibly in preparation for the upcoming fall migration (George *et al.*, 1987).

If such changes in the muscles of waterfowl do not result from use/disuse, what is the mechanism that regulates muscle change? During the molting period, changes occur in plasma concentrations of several hormones. High concentrations of thyroid hormone (T_3 and T_4) and a decrease in the concentration of testosterone are thought to stimulate the onset of molt (Assenmacher *et al.*, 1975), and growth hormone is at its peak concentration in the plasma of molting waterfowl (Scanes *et al.*, 1980). It is possible that the changes in muscle could be regulated, at least in part, through changes in these hormones, as each has been shown to alter various properties of muscle.

Although the mechanism(s) that regulate rapid atrophy and hypertrophy in the muscle of waterfowl are unknown, such an adaptation for nonuse hypertrophy could be beneficial for migratory waterfowl. Rapid hypertrophy of the flight muscles would ensure that these birds would have sufficient power to lift off the ground before the flight feathers are fully regrown. Once waterfowl can achieve flight, then endurance training may further enhance the capabilities of the flight muscles (Butler and Turner, 1988).

B. Seasonal Changes in Muscle Biochemical Properties

Seasonal changes in the biochemical properties of the flight muscles in birds also have been reported. Lundgren and Kiessling (1985) noted changes in the oxidative enzymes 3-hydroxyacyl CoA dehydrogenase, citrate synthase, and cytochrome oxidase between migratory and nonmigratory periods for several bird species. The activities of all three enzymes were greatest during the migratory period and lowest during the breeding season. In contrast, the activities of the glycolytic enzymes 6-phosphofructokinase and lactate dehydrogenase were lowest during the migratory period and highest during the breeding period. Lundgren and Kiessling (1985) suggested that seasonal changes in enzyme activities were the result of changes in locomotor activities of the birds. On the other hand, Marsh (1981) failed to find increased specific activity of citrate synthase in the *pectoralis* muscle of the Gray catbird in either the migratory or premigratory period, although the mass of the flight muscle increased.

The studies of both Marsh (1981) and Lundgren and Kiessling

(1985) showed increases in the total aerobic capacities of the flight muscles during the migratory period. An increase in the total aerobic capacity of the *pectoralis* muscles would likely increase the endurance capacity of this muscle by allowing it to operate below its maximal rate of oxygen consumption (Marsh, 1981). Such an adaptation would be invaluable for the upcoming migratory flights.

It is interesting that migratory waterfowl apparently have the ability to either maintain or rapidly regain the aerobic potential of the flight muscles during their flightless period. Both the normal and atrophied *pectoralis* muscles of Eared grebes were found to contain high densities of mitochondria (Gaunt *et al.*, 1990). The maintenance of high mitochondrial densities in the atrophied flight muscle was suggested to provide a "jump start" for rapid muscle regrowth, as in the recovery process of the flight muscles of the grebes. This rapid recovery of flight muscle mass did not appear to require either prolonged or high-intensity exercise (Gaunt *et al.*, 1990). Thus, such seasonal changes may be an anticipatory response for the upcoming migration of these birds.

The concentration of myoglobin in the *pectoralis* muscle of birds also appears to vary between migratory and nonmigratory periods. Significantly higher concentrations of myoglobin have been found in the heart and *pectoralis* muscles of Starlings upon their arrival from migration than in Starlings that had been resting on their wintering grounds (Palacios *et al.*, 1984). This increased myoglobin concentration has been suggested to result from increased use of the flight and cardiac muscles during migration. Whether myoglobin, like citrate synthase and muscle mass, increases with little or no exercise in birds is currently unknown.

VI. Summary

Birds are excellent endurance athletes. Not only do many birds undertake long migratory flights, but many do so under extreme environmental conditions: excessive heat, extreme cold, and the hypoxic conditions of high altitude. We are just now starting to understand the physiological adaptations these animals possess for surviving and thriving in these environments. Still, relatively few studies have actually been performed on exercising birds, particularly on birds flying under the conditions mentioned here. Furthermore, not all birds are capable of sustained exercise in hypoxia, heat, and cold. More work is needed to increase our understanding of the differences in the physiological systems that allow some birds to be better able to exercise under such conditions.

References

Abdalla, M. A., Maina, J. N., King, A. S., King, D. Z., and Henry, J. (1982). Morphometrics of the avian lung. 1. The domestic fowl (*Gallus gallus* variant *domesticus*). *Respir. Physiol.* **47,** 267–278.

Åkeson, Å., Ehrenstein, G. V., Hévésy, G., and Theorell, H. (1960). Life span of myoglobin. *Arch. Biochem. Biophys.* **91,** 310–318.

Alp, P. R., Newsholme, E. S., and Zammit, V. A. (1976). Activities of citrate synthase and NAD^+-linked and $NADP^+$-linked isocitrate dehydrogenase in muscle from vertebrates and invertebrates. *Biochem. J.* **154,** 689–700.

Andersen, P., and Henricksson, J. (1977). Capillary supply of the quadriceps femoris muscle of man: Adaptive response to exercise. *J. Physiol. (London)* **220,** 677–690.

Anderson, H. T., Hustvedt, B. E., and Lovo, A. (1965). Acid–base changes in diving ducks. *Acta Physiol. Scand.* **63,** 128–132.

Ankney, C. D. (1979). Does the wing molt cause nutritional stress in lesser snow geese? *Auk* **96,** 68–72.

Ankney, C. D. (1984). Nutrient reserve dynamics of breeding and molting brant. *Auk* **101,** 361–370.

Assenmacher, I., Astier, H., Daniel, J. Y., and Jallageas, M. (1975). Experimental studies on the annual cycles of thyroid and adrenocortical functions in relation to the reproductive cycle of drakes. *J. Physiol. (Paris)* **70,** 507–520.

Åstrand, P.-O., Cuddy, T. E., Saltin, B., and Stenberg, J. (1964). Cardiac output during submaximal and maximal work. *J. Appl. Physiol.* **19,** 268–274.

Aulie, A. (1971). Co-ordination between the activity of the heart and the flight muscles during flight in small birds. *Comp. Biochem. Physiol. A* **38A,** 91–97.

Aulie, A. (1972). Co-ordination between the activity of the heart and the pectoral muscles during flight in the pigeon. *Comp. Biochem. Physiol. A* **41A,** 43–48.

Bailey, R. O. (1985). Protein reserve dynamics in postbreeding adult male redheads. *Condor* **87,** 23–32.

Baldwin, J., Jardel, J.-P., Montague, T., and Tomkin, R. (1984). Energy metabolism in penguin swimming muscles. *Mol. Physiol.* **6,** 33–42.

Banzett, R. B., Nations, C. S., Wang, N., Butler, J. P., and Lehr, J. L. (1992). Mechanical independence of wingbeat and breathing in starlings. *Respir. Physiol.* **89,** 27–36.

Barger, A. C., Richards, V., Metcalfe, J., and Gunther, B. (1956). Regulation of the circulation during exercise. Cardiac output (direct Fick) and metabolic adjustments in the normal dog. *Am. J. Physiol.* **184,** 613–623.

Barnas, G. M., and Rautenberg, W. (1984). Respiratory responses to shivering produced by external and central cooling in the pigeon. *Pflüegers Arch.* **401,** 228–232.

Barnas, G. M., and Rautenberg, W. (1990). Shivering and cardiorespiratory responses during normocapnic hypoxia in the pigeon. *J. Appl. Physiol.* **68,** 84–87.

Bartholomew, G. A., and Cade, T. J. (1956). Water consumption of house finches. *Condor* **58,** 406–412.

Baudinette, R. V., and Gill, P. (1985). The energetics of "flying" and "paddling" in water: Locomotion in penguins and ducks. *J. Comp. Physiol.* **155,** 373–380.

Baudinette, R. V., Loveridge, J. P., Wilson, K. J., Mills, C. D., and Schmidt-Nielsen, K.

(1976). Heat loss from the feet of herring gulls at rest and during flight. *Am. J. Physiol.* **230,** 920–924.
Baudinette, R. V., Tonkin, A. L., Orbach, J., Seymour, R. S., and Wheldrake, J. F. (1982). Cardiovascular function during treadmill exercise in the turkey. *Comp. Biochem. Physiol. A* **72A,** 327–332.
Bauer, C., and Petschow, D. (1976). Functional studies on hemoglobin from highland mammals and birds. *Physiologist* **19,** 119.
Bech, C., and Nomoto, S. (1982). Cardiovascular changes associated with treadmill running in the Pekin duck. *J. Exp. Biol.* **97,** 345–358.
Bech, C., Johansen, K., Brent, R., and Nicol, S. (1984). Ventilatory and circulatory changes during cold exposure in the Pekin duck, *Anas platyrhynchos. Respir. Physiol.* **57,** 103–112.
Bech, C., Rautenberg, W., and May, B. (1985). Ventilatory oxygen extraction during cold exposure in the pigeon (*Columba livia*). *J. Exp. Biol.* **116,** 499–502.
Bellrose, F. C. (1976). "Ducks, Geese and Swans of North America." Stackpole Books, Harrisburgh, PA.
Berger, M., Hart, J. S., and Roy, O. Z. (1970). Respiration, oxygen consumption and heart rate in some birds during rest and flight. *Z. Vergl. Physiol.* **66,** 201–214.
Berger, M., Hart, J. S., and Roy, O. Z. (1971). Respiratory water and heat loss of the black duck during flight at different ambient temperatures. *Can. J. Zool.* **49,** 767–774.
Bernstein, M. H. (1976). Ventilation and respiratory evaporation in the flying crow *Corvus ossifragus. Respir. Physiol.* **26,** 371–382.
Bernstein, M. H. (1987). Respiration in flying birds. *In* "Bird Respiration" (T. J. Seller, ed.), pp. 43–73. CRC Press, Boca Raton, FL.
Bernstein, M. H. (1989). Body and brain temperatures in pigeons at simulated high altitudes. *In* "Physiology of Cold Adaptation in Birds" (C. Bech and R. E. Reinertsen, eds.), pp. 207–210. Plenum, New York.
Biesel, W., and Nachtigall, W. (1987). Pigeon flight in a wind tunnel. IV. Thermoregulation and water homeostasis. *J. Comp. Physiol. B* **157B,** 117–128.
Black, C. P., and Tenney S. M. (1980a). Oxygen transport during progressive hypoxia in high altitude and sea-level waterfowl. *Respir. Physiol.* **39,** 217–239.
Black, C. P., and Tenney S. M. (1980b). Pulmonary hemodynamics responses to acute and chronic hypoxia in two waterfowl species. *Comp. Biochem. Physiol. A* **67A,** 291–293.
Black, C. P., Tenney, S. M., and Van Kroonenburg, M. (1978). Oxygen transport during progressive hypoxia in bar-headed geese (*Anser indicus*) acclimatized to sea level and 5600 meters. *In* "Respiratory Function in Birds, Adult and Embryonic" (J. Piiper, ed.), pp. 79–83. Springer-Verlag, Berlin.
Blix, A. S., and Folkow, B. (1983). Cardiovascular adjustments to diving in mammals and birds. *In* "Handbook of Physiology" (J. T. Shepherd and F. M. Abboud, eds.), Sect. 2, Vol. III, pp. 917–945. Am. Physiol. Soc., Bethesda, MD.
Blix, A. S., and Berg, T. (1974). Arterial hypoxia and the diving responses of ducks. *Acta Physiol. Scand.* **92,** 566–568.
Boggs, D. F., and Birchard, G. F. (1982). Relationship between hypoxic ventilatory response and hemoglobin O_2-affinity in birds. *Physiologist* **25,** 325.
Bouverot, P., Hildwein, G., and Oulhen, P. H. (1976). Ventilatory and circulatory O_2 convection at 4000 m in pigeon at neutral and cold temperature. *Respir. Physiol.* **28,** 371–385.
Brackenbury, J. H., and Williamson, A. D. B. (1989). Treadmill exercise training increases the oxidative capacity of chicken iliotibialis muscle. *Poult. Sci.* **68,** 577–581.
Brent, R., Rasmussen, J. G., Bech, C., and Martini, S. (1983). Temperature dependence of

ventilation and O_2-extraction in the kittiwake, *Rissa tridactyla. Experientia* **39,** 1092–1093.
Brent, R., Pedersen, P. F., Bech, C., and Johansen, K. (1984). Lung ventilation and temperature regulation in the European coot, *Fulica atra. Physiol. Zool.* **57,** 19–25.
Brown, L. (1976). "Birds of Prey, Their Biology and Ecology." Hamlyn, London.
Bryan, R. M., and Jones, D. R. (1980). Cerebral energy metabolism in diving and non-diving birds during hypoxia and apnoeic asphyxia. *J. Physiol. (London)* **299,** 323–336.
Bucher, T. L. (1981). Oxygen consumption, ventilation and respiratory heat loss in a parrot, *Bolborhynchus lineola*, in relation to ambient temperature. *J. Comp. Physiol.* **142,** 479–488.
Burton, H. W., and Barclay, J. K. (1986). Metabolic factors from exercising muscle and the proliferation of endothelial cells. *Med. Sci. Sports Exercise* **18,** 390–395.
Butler, P. J. (1991). Exercise in birds. *J. Exp. Biol.* **160,** 233–262.
Butler, P. J., and Jones, D. R. (1971). The effect of variations in heart rate and regional distribution of blood flow on the normal pressor response to diving in ducks. *J. Physiol. (London)* **214,** 457–479.
Butler, P. J., and Jones, D. R. (1982). The comparative physiology of diving in vertebrates. *Adv. Comp. Physiol. Biochem.* **8,** 179–364.
Butler, P. J., and Turner, D. L. (1988). Effect of training on maximal oxygen uptake and aerobic capacity of locomotory muscles in tufted ducks (*Aythya fuligula*). *J. Physiol. (London)* **401,** 347–359.
Butler, P. J., and Woakes, A. J. (1976). Changes in heart rate and respiratory frequency associated with natural submersion in ducks. *J. Physiol. (London)* **256,** 73–74.
Butler, P. J., and Woakes, A. J. (1979). Changes in heart rate and respiratory frequency during natural behaviour of ducks, with particular reference to diving. *J. Exp. Biol.* **79,** 283–300.
Butler, P. J., and Woakes, A. J. (1980). Heart rate, respiratory frequency, and wing beat frequency of free flying barnacle geese, *Branta leucopsis. J. Exp. Biol.* **85,** 213–226.
Butler, P. J., and Woakes, A. J. (1982). Telemetry of physiological variables from diving and flying birds. *Symp. Zool. Soc. London* **49,** 107–128.
Butler, P. J., and Woakes, A. J. (1984). Heart rate and aerobic metabolism in Humboldt penguins *Spheniscus humboldti* during voluntary dives. *J. Exp. Biol.* **108,** 419–428.
Butler, P. J., West, N. H., and Jones, D. R. (1977). Respiratory and cardiovascular responses of the pigeon to sustained level flight in a wind tunnel. *J. Exp. Biol.* **71,** 7–26.
Butler, P. J., Turner D. L., Al-Wassia, A., and Bevan, R. M. (1988). Regional distribution of blood flow during swimming in the tufted duck (*Aythya fuligula*). *J. Exp. Biol.* **135,** 461–472.
Catlett, R. H., Walters, T. W., and Dutro, P. A. (1978). The effect of flying and not flying on myoglobin content of heart muscle of the pigeon *Columba livia domestica. Comp. Biochem. Physiol. A* **59A,** 401–402.
Christensen, E. H., and Dill, D. B. (1935). Oxygen dissociation curves of bird blood. *J. Biol. Chem.* **109,** 443–448.
Clausen, G., Sanson, R., and Storesund, A. (1971). The HbO_2 dissociation curve of the fulmar and the herring gull. *Respir. Physiol.* **12,** 66–70.
Clemens, D. L. (1990). Interspecific variation and effects of altitude on blood properties of rosy finches (*Leucosticte arctoa*) and house finches (*Carpodacus mexicanus*). *Physiol. Zool.* **63,** 288–307.
Cohn, J. E., Krog, J., and Shannon, R. (1968). Cardiopulmonary responses to head immersion in domestic geese. *J. Appl. Physiol.* **25,** 36–41.

Covell, D. G., and Jacquez, J. A. (1987). Does myoglobin contribute significantly to diffusion of oxygen in red skeletal muscle? *Am. J. Physiol.* **252,** R341–R347.

Davies, D. G., Nolan, W. F., and Sexton, J. A. (1986). Effect of hypocapnia on ventral medullary blood flow and pH during hypoxia in cats. *J. Appl. Physiol.* **61,** 87–90.

Davis, M. B., and Guderley, H. (1987). Energy metabolism in the locomotor muscles of the common murre (*Uria aalge*) and the Atlantic puffin (*Fratercula arctica*). *Auk* **104,** 733–739.

Davis, M. B., and Guderley, H. (1990). Biochemical adaptations to diving in the common murre *Uria aalge* and the Atlantic puffin, *Fratercula arctica*. *J. Exp. Zool.* **253,** 235–244.

Dawson, W. R. (1982). Evaporative losses of water by birds. *Comp. Biochem. Physiol. A* **71A,** 495–509.

Dubach, M. (1981). Quantitative analysis of the respiratory system of the house sparrow, budgerigar and violet-eared hummingbird. *Respir. Physiol.* **46,** 43–60.

DuBowy, P. J. (1985). Seasonal organ dynamics in post-breeding male blue-winged teal and northern shovelers. *Comp. Biochem. Physiol. A* **82A,** 899–906.

Duncker, H.-R. (1973). Der quantitative Aufbau des Lungen-Luftsacksystems der Vogel. *Verh. Anat. Ges.* **67,** 197–204.

Enoki, Y., Morimoto, T., Nakatani, A., Sakata, S., Ohga, Y., Kohzuki, H., and Shimizu, S. (1987). Wide variation of myoglobin contents in gizzard smooth muscles of various avian species. *In* "Oxygen Transport to Tissues X" (M. Mochizuki, C. R. Honig, T. Koyama, T. K. Goldstick, and D. F. Bruley, eds.), pp. 709–715. Plenum, New York.

Faraci, F. M. (1988). Oxygen transport in birds during hypoxia. *Ann. Sports Med.* **4,** 178–183.

Faraci, F. M. (1991). Adaptations to hypoxia in birds: How to fly high. *Annu. Rev. Physiol.* **53,** 59–70.

Faraci, F. M., and Fedde, M. R. (1986). Regional circulatory responses to hypocapnia and hypercapnia in bar-headed geese. *Am. J. Physiol.* **250,** R499–R504.

Faraci, F. M., Kilgore, D. L., and Fedde, M. R. (1984a). Oxygen delivery to the heart and brain during hypoxia: Pekin duck *vs.* bar-headed goose. *Am. J. Physiol.* **247,** R69–R75.

Faraci, F. M., Kilgore, D. L., and Fedde, M. R. (1984b). Attenuated pulmonary pressor response to hypoxia in bar-headed geese. *Am. J. Physiol.* **247,** R402–R403.

Faraci, F. M., Kilgore, D. L., and Fedde, M. R. (1985). Blood flow distribution during hypocapnic hypoxia in Pekin ducks and bar-headed geese. *Respir. Physiol.* **61,** 21–30.

Fedak, M. A., and Seeherman, J. H. (1979). Reappraisal of energetics of locomotion shows identical cost in bipeds and quadrupeds including ostrich and horse. *Nature (London)* **282,** 713–716.

Fedde, M. R., Faraci, F. M., Kilgore, D. L., Cardinet, G. H., and Chatterjee, A. (1985). Cardiopulmonary adaptations in birds for exercise at high altitude. *In* "Circulation, Respiration and Metabolism: Current Comparative Approaches" (R. Gilles, ed.), pp. 148–162. Springer-Verlag, New York.

Fedde, M. R., Orr, J. A., Shams, H., and Scheid, P. (1989). Cardiopulmonary function in exercising bar-headed geese during normoxia and hypoxia. *Respir. Physiol.* **77,** 239–262.

Folkow, B., Fuxe, K., and Sonnenschein, R. R. (1966). Responses of skeletal musculature and its vasculature during "diving" in the duck: Peculiarities of the adrenergic vasoconstrictor innervation. *Acta Physiol. Scand.* **67,** 327–342.

Fry, C. H., Ash, J. S., and Ferguson-Lees, I. J. (1970). Spring weights of some palearctic migrants at Lake Chad. *Ibis* **112,** 58–82.

Fry, C. H., Ferguson-Lees, I. J., and Dowsett, R. J. (1972). Flight muscle hypertrophy and

ecophysiological variation of yellow wagtail *Moticilla flava* races at Lake Chad. *J. Zool.* **167,** 293–306.
Gaehtgens, P. (1990). Avian versus mammalian blood rheology: Experimental observations and physiological relevance. *In* "Hypoxia: The Adaptations" (J. R. Sutton, G. Coates, and J. E. Remmers, eds.), pp. 20–25. B. C. Decker, Toronto.
Gaehtgens, P., Schmidt, F., and Will, G. (1981a). Comparative rheology of nucleated and non-nucleated red blood cells. I. Microrheology of avian erythrocytes during capillary flow. *Pflüegers Arch.* **390,** 278–282.
Gaehtgens, P., Schmidt, F., and Will, G. (1981b). Comparative rheology of nucleated and non-nucleated red blood cells. II. Rheological properties of avian red cell suspensions in narrow capillaries. *Pflüegers Arch.* **390,** 283–287.
Gaunt, A. S., Hikida, R. S., Jehl, J. R., and Fenbert, L. (1990). Rapid atrophy and hypertrophy of an avian flight muscle. *Auk* **107,** 649–659.
Gayeski, T. E. J., and Honig, C. R. (1986). O_2 gradients from sarcolemma to cell interior in red muscle at maximal $\dot{V}_{O_2}$. *Am. J. Physiol.* **251,** H789–H799.
Gehr, P., Mwangi, D. K., Ammann, A., Maloiy, G. M. O., Taylor, C. R., and Weibel, E. R. (1981). Design of the mammalian respiratory system. V. Scaling morphometric pulmonary diffusing capacity to body mass: Wild and domestic mammals. *Respir. Physiol.* **44,** 61–86.
Gentry, R. L., and Kooyman, G. L. (1986). "Fur Seals, Maternal Strategies on Land and at Sea." Princeton Univ. Press, Princeton, NJ.
George, J. C., and Berger, A. J. (1966). "Avian Myology." Academic Press, New York.
George, J. C., John, T. M., and Minhas, K. J. (1987). Seasonal degradative, reparative and regenerative ultrastructural changes in the breast muscle of the migratory Canada goose. *Cytobios* **52,** 109–126.
Gessaman, J. A. (1980). An evaluation of heart rate as an indirect measure of daily energy metabolism of the American kestrel. *Comp. Biochem. Physiol. A* **65A,** 273–289.
Gleeson, M., and Brackenbury, J. H. (1983). Respiratory and blood gas responses in exercising birds. *Comp. Biochem. Physiol. A* **76A,** 211–216.
Greenewalt, C. H. (1962). Dimensional relationships for flying animals. *Smithson. Misc. Collect.* **144,** 1–46.
Greenwald, L., Stone, W. B., and Cade, T. J. (1967). Physiological adjustments of the budgerygah (*Melopsittacus undulatus*) to dehydrating conditions. *Comp. Biochem. Physiol.* **22,** 91–100.
Grubb, B. R. (1982). Cardiac output and stroke volume in exercising ducks and pigeons. *J. Appl. Physiol.* **53,** 207–211.
Grubb, B. R. (1983). Allometric relations of cardiovascular function in birds. *Am. J. Physiol.* **245,** H567–H572.
Grubb, B. R., Mills, C. D., Colacino, J. M., and Schmidt-Nielsen, K. (1977). Effect of arterial carbon dioxide on cerebral blood flow in ducks. *Am. J. Physiol.* **232,** H596–H601.
Grubb, B. R., Jones, J. H., and Schmidt-Nielsen, K. (1979). Avian cerebral blood flow: Influence of the Bohr effect on oxygen supply. *Am. J. Physiol.* **236,** H744–H749.
Grubb, B. R., Jorgensen, D. D., and Conner, M. (1983). Cardiovascular changes in the exercising emu. *J. Exp. Biol.* **104,** 193–201.
Haggblom, L., Terwilliger, R. C., and Terwilliger, N. B. (1988). Changes in myoglobin and lactate dehydrogenase in muscle tissues of a diving bird, the pigeon guillemot, during maturation. *Comp. Biochem Physiol. B* **91B,** 273–277.
Hagler, L., Coppes, R. I., Jr., Askew, E. W., Hecker, A. L., and Herman, R. H. (1980). The

influence of exercise and diet on myoglobin and metmyoglobin reductase in the rat. *J. Lab. Clin. Med.* **95,** 222–230.
Hart, J. S., and Roy, O. Z. (1966). Respiratory and cardiac response to flight in pigeons. *Physiol. Zool.* **39,** 291–306.
Hart, J. S., and Roy, O. Z. (1967). Temperature regulation during flight in pigeons. *Am. J. Physiol.* **213,** 1311–1316.
Harvey, E. N., McElroy, W. D., Whiteley, A. H., Warren, G. H., and Pease, D. C. (1944). Bubble formation in animals. III. An analysis of gas tension and hydrostatic pressure in cats. *J. Cell. Comp. Physiol.* **24,** 117–132.
Henshaw, H. W. (1910). Migration of the Pacific plover to and from the Hawaiian Islands. *Auk* **27,** 245–262.
Hickson, R. C. (1981). Skeletal muscle cytochrome C and myoglobin, endurance and frequency of training. *J. Appl. Physiol.* **51,** 746–749.
Hirth, K.-D., Biesel, W., and Nachtigall, W. (1987). Pigeon flight in a wind tunnel. III. Regulation of body temperature. *J. Comp. Physiol. B* **157B,** 111–116.
Holle, J. P., Meyer, M., and Scheid, P. (1977). Oxygen affinity of duck blood determined by *in vivo* and *in vitro* technique. *Respir. Physiol.* **29,** 355–361.
Holt, J. P., Rhode, E. A., and Kines, H. (1968). Ventricular volumes and body weight in mammals. *Am. J. Physiol.* **215,** 704–715.
Hudlicka, O., Wright, A. J. A., and Ziada, A. M. A. R. (1984). Capillary growth in the heart. *Prog. Appl. Microcirc.* **4,** 16–27.
Hudson, D. M., and Bernstein, M. H. (1981). Temperature regulation and heat balance in flying white-necked ravens, *Corvus cryptoleucus*. *J. Exp. Biol.* **90,** 267–281.
Hurtado, A., Rotta, A., Merino, C., and Pons, J. (1937). Studies of myohemoglobin at high altitudes. *Am. J. Med. Sci.* **194,** 708–713.
Isaacks, R. E. (1985). The relationship between erythrocyte phosphate metabolism, carbon dioxide and pH on blood oxygen affinity in birds. *In* “Circulation, Respiration and Metabolism: Current Comparative Approaches” (R. Gilles, ed.), pp. 301–311. Springer-Verlag, New York.
Johannesen, H., and Nicol, S. C. (1990). Effects of cold exposure on oxygen consumption, ventilation and interclavicular air-sac gases in the little penguin (*Eudyptula minor*). *J. Exp. Biol.* **154,** 397–405.
Johansen, K. (1964). Regional distribution of circulating blood during submersion asphyxia in the duck. *Acta Physiol. Scand.* **62,** 1–9.
Johansen, K., Berger, M., Bicudo, J. E. P. W., Ruschi, A., and De Almeida, P. J. (1987). Respiratory properties of blood and myoglobin in hummingbirds. *Physiol. Zool.* **60,** 269–278.
Johnston, D. W., and McFarlane, R. W. (1967). Migration and bioenergetics of flight in the Pacific golden plover. *Condor* **69,** 156–168.
Jones, D. R., Bryan, R. M., West, N. H., Lord, R. H., and Clark, B. (1979). Regional distribution of blood flow during diving in the duck (*Anas platyrhynchos*). *Can. J. Zool.* **57,** 995–1002.
Jones, J. H., and Lindstedt, S. L. (1993). Limits to maximal performance. *Annu. Rev. Physiol.* **55,** 547–569.
Kaiser, C. E., and George, J. C. (1973). Interrelationship amongst the avian orders Galliformes, Columbiformes and Anseriformes as evinced by the fiber types in the pectoralis muscle. *Can. J. Zool.* **51,** 887–892.
Kanwisher, J. W., Gabrielsen, G., and Kanwisher, N. (1981). Free and forced diving in birds. *Science* **211,** 717–719.
Kiley, J. P., Kuhlmann, W. D., and Fedde, M. R. (1982). Ventilatory and blood gas adjustments in exercising isothermic ducks. *J. Comp. Physiol.* **147,** 107–112.

Kiley, J. P., Faraci, F. M., and Fedde, M. R. (1985). Gas exchange during exercise in hypoxic ducks. *Respir. Physiol.* **59,** 105–115.
King, C. E., Dodd, S. L., and Cain, S. M. (1987). O_2 delivery to contracting muscle during hypoxic or CO hypoxia. *J. Appl. Physiol.* **63,** 726–732.
King, J. R., and Farner, D. S. (1963). The relationship of fat deposition to zugunruhe and migration. *Condor* **65,** 200–223.
Klausen, K., Anderson, L. B., and Pelle, I. (1981). Adaptive changes in work capacity, skeletal muscle capillarization and enzyme levels during training and detraining. *Acta Physiol. Scand.* **113,** 9–16.
Kooyman, G. L. (1975). Behaviour and physiology of diving. *In* "The Biology of Penguins" (B. Stonehouse, ed.), pp. 115–137. Macmillan, London.
Kooyman, G. L. (1981). "Weddell Seal, Consummate Diver." Cambridge Univ. Press, Cambridge, UK.
Kooyman, G. L., and Ponganis, P. J. (1990). Behavior and physiology of diving in emperor and king penguins. *In* "Penguin Biology" (L. S. Davis and J. T. Darby, eds.), pp. 229–242. Academic Press, San Diego.
Kooyman, G. L., Drabek, C. M., Elsner, R., and Campbell, C. M. (1971). Diving behavior of the emperor penguin, *Aptenodytes forsteri. Auk* **88,** 775–795.
Kooyman, G. L., Schroeder, J. P., Greene, D. G., and Smith, V. A. (1973). Gas exchange in penguins during simulated dives to 30 and 68 m. *Am. J. Physiol.* **225,** 1467–1471.
Kooyman, G. L., Davis, R. W., Croxall, J. P., and Costa, D. P. (1982). Diving depths and energy requirements of king penguins. *Science* **217,** 726–727.
Lawrie, R. A. (1953). The activity of the cytochrome system in muscle and its relation to myoglobin. *Biochem. J.* **55,** 298–305.
Laybourne, R. C. (1974). Collision between a vulture and an aircraft at an altitude of 37,000 feet. *Wilson Bull.* **86,** 461–462.
Lenfant, C., Kooyman, G. L., Elsner, R., and Drabek, C. M. (1969). Respiratory function of blood of the Adelie penguin *Pygoscelis adeliae. Am. J. Physiol.* **216,** 1598–1600.
Lierse, V. W. (1963). Die Kappillardichte im wirbeltiegehirn. *Acta Anat.* **54,** 1–31.
Lord, R. D., Bellrose, F. C., and Cochran, W. W. (1962). Radiotelemetry of the respiration of a flying duck. *Science* **137,** 39–40.
Lundgren, B. O., and Kiessling, K.-H. (1985). Seasonal variation in catabolic enzyme activities in breast muscle of some migratory birds. *Oecologia* **66,** 468–471.
Lundgren, B. O., and Kiessling, K.-H. (1988). Comparative aspects of fiber types, areas and capillary supply in the pectoralis muscle of some passerine birds with differing migratory behavior. *J. Comp. Physiol. B* **158B,** 165–173.
Lutz, P. L., Longmuir, I. S., and Schmidt-Nielsen, K. (1974). Oxygen affinity of bird blood. *Respir. Physiol.* **20,** 325–330.
Maginniss, L. A. (1985). Blood oxygen transport in the house sparrow, *Passer domesticus. J. Comp. Physiol. B* **155B,** 277–283.
Maina, J. N. (1984). Morphometrics of the avian lung. 3. The structural design of the passerine lung. *Respir. Physiol.* **55,** 291–307.
Maina, J. N., and King, A. S. (1982). Morphometrics of the avian lung. 2. The wild mallard (*Anas platyrhynchos*) and graylag goose (*Anser anser*). *Respir. Physiol.* **50,** 299–310.
Manville, R. H. (1963). Altitude record for mallard. *Wilson Bull.* **75,** 92.
Marsh, R. L. (1981). Catabolic enzyme activities in relation to premigratory fattening and muscle hypertrophy in the gray catbird (*Dumetella carolinensis*). *J. Comp. Physiol. B* **141B,** 417–423.
Marsh, R. L. (1984). Adaptations of the grey catbird *Dumetella carolinensis* to long-

distance migration: Flight muscle hypertrophy associated with elevated body mass. *Physiol. Zool.* **57,** 105–117.
Marsh, R. L., and Dawson, W. R. (1982). Substrate metabolism in seasonally acclimatized American goldfinches. *Am. J. Physiol.* **242,** R563–R569.
Marsh, R. L., and Storer, R. W. (1981). Correlation of flight-muscle size and body mass in Cooper's hawks: A natural analogue of power training. *J. Exp. Biol.* **91,** 363–368.
Mathieu-Costello, O. (1990). Histology of flight: Tissue and muscle gas exchange. *In* "Hypoxia: The Adaptations" (J. R. Sutton, G. Coates, and J. E. Remmers, eds.), pp. 13–19. B. C. Decker, Toronto.
Meinertzhagen, R. (1955). The speed and altitude of bird flight. *Ibis* **97,** 81–117.
Mill, G. K., and Baldwin, J. (1983). Biochemical correlates of swimming and diving behavior in the little penguin *Eudyptula minor. Physiol. Zool.* **56,** 242–254.
Millard, R. W., Johansen, K., and Milsom, W. K. (1973). Radiotelemetry of cardiovascular responses to exercise and diving in penguins. *Comp. Biochem. Physiol. A* **46A,** 227–240.
Monge, C., and Leon-Velarde, F. (1991). Physiological adaptation to high altitude: Oxygen transport in mammals and birds. *Physiol. Zool.* **71,** 1135–1172.
Moreau, R. E. (1961). Problems of Mediterranean–Saharan migrations. *Ibis* **103,** 373–427.
Morgan, V. E., and Chichester, D. F. (1935). Properties of the blood of the domestic fowl. *J. Biol. Chem.* **110,** 285–298.
Murphy, B. J., Zapol, W. M., and Hochachka, P. W. (1980). Metabolic activities of heart, lung and brain during diving and recovery in the Weddell seal. *J. Appl. Physiol.* **48,** 596–605.
Murrish, D. E. (1970). Responses to diving in the dipper *Cinclus mexicanus. Comp. Biochem. Physiol.* **34,** 853–858.
Murry, J. F., Gold, P., and Johnson, B. L. (1963). The circulatory effects of hematocrit variations in normovolemic and hypervolemic dogs. *J. Clin. Invest.* **42,** 139–143.
Nagy, K. A., Siegfried, W. R., and Wilson, R. P. (1984). Energy utilization by free-ranging Jackass penguins, *Spheniscus demersus. Ecology* **65,** 1648–1655.
Nisbet, I. C. T., Drury, W. H., and Baird, J. (1963). Weight-loss during migration. I. Deposition and consumption c 'fat by the blackpoll warbler *Dendroica striata. Bird Banding* **34,** 107–138.
Nishida, J. (1976). Changes in myoglobin content during development and growth of chicken. *Jpn. J. Vet. Sci.* **38,** 299–303.
Nishida, J., and Nishida, T. (1978). Immunological and quantitative comparison of myoglobin components in various muscle tissues of the chicken (*Gallus gallus domesticus*). *Jpn. J. Vet Sci.* **40,** 385–399.
Odum, E. P., Connell, C. E., and Stoddard, H. L. (1961). Flight energy and estimated flight ranges of some migratory birds. *Auk* **78,** 515–527.
Pages, T., and Planas, J. (1983). Muscle myoglobin and flying habits in birds. *Comp. Biochem. Physiol. A* **74A,** 289–294.
Palacios, L., Palomeque, J., Riera, M., Pages, T., Viscor, G., and Planas, J. (1984). Oxygen transport properties in the starling, *Sturnus vulgaris* L. *Comp. Biochem. Physiol. A* **77A,** 255–260.
Parker, G. H., and George, J. C. (1973). Glycogen utilization by the white fibers in the pigeon pectoralis as main energy process during shivering thermogenesis. *Comp. Biochem. Physiol. A* **50A,** 433–437.
Parker, G. H., and George, J. C. (1976). Effects of intense exercise on intracellular glycogen and fat in pigeon pectoralis. *Acta Anat.* **96,** 568–573.

Pattengale, P. K., and Holloszy, J. O. (1967). Augmentation of skeletal muscle myoglobin by a program of treadmill running. *Am. J. Physiol.* **213,** 783–785.

Paulev, P. (1965). Decompression sickness following repeated breath-hold dives. *J. Appl. Physiol.* **20,** 1028–1031.

Pennyquick, C. J. (1975). Mechanics of flight. *Avian Biol.* **5,** 1–75.

Piatt, J. F., and Nettleship, D. N. (1985). Diving depths of four alcids. *Auk* **102,** 293–297.

Piiper, J., and Scheid, P. (1975). Gas transport efficacy of gills, lungs and skin: Theory and experimental data. *Respir. Physiol.* **23,** 209–221.

Pinshow, B., Bernstein, M. H., and Arad, Z. (1985). Effects of temperature and P_{CO_2} on O_2 affinity of pigeon blood: Implications for brain O_2 supply. *Am. J. Physiol.* **249,** R758–R764.

Powell, F. L. (1982). Diffusion in avian lungs. *Fed. Proc., Fed. Am. Soc. Exp. Biol.* **41,** 2131–2133.

Powell, F. L., and Mazzone, R. W. (1983). Morphometrics of rapidly frozen goose lungs. *Respir. Physiol.* **51,** 319–332.

Powell, F. L., and Wagner, P. D. (1982). Ventilation–perfusion inequality in avian lungs. *Respir. Physiol.* **48,** 233–241.

Prange, H. D., and Schmidt-Nielsen, K. (1970). The metabolic cost of swimming in ducks. *J. Exp. Biol.* **53,** 763–777.

Rakusan, K., Ost'adal, B., and Wachtlova, M. (1971). The influence of muscular work on the capillary density in the heart and skeletal muscle of pigeon (*Columba livia*). *Can. J. Physiol. Pharmacol.* **49,** 167–170.

Raveling, D. G. (1979). The annual cycle of body composition of Canada geese with special reference to control of reproduction. *Auk* **96,** 234–252.

Reynafarje, C., Faura, J., Villavicencio, D., Curaca, A., Reynafarje, B., Oyola, L., Contreras, L., Vallenas, E., and Faura, A. (1975). Oxygen transport of hemoglobin in high-altitude animals (Camelidae). *J. Appl. Physiol.* **38,** 806–810.

Rosser, B. W. C., and George, J. C. (1986). The avian pectoralis: Histochemical characterization and distribution of muscle fiber types. *Can. J. Zool.* **64,** 1174–1185.

Rosser, B. W. C., and George, J. C. (1987). Ultrastructural and cytological changes in the muscle fibers of the pectoralis of the giant Canada goose (*Branta canadensis maxima*) in disuse atrophy during molt. *Cell Tissue Res.* **247,** 689–696.

Sarelius, I. H. (1989). Microcirculation in striated muscle after acute reduction in systemic hematocrit. *Respir. Physiol.* **78,** 7–17.

Saunders, D. K., and Fedde, M. R. (1991). Physical conditioning: Effect on the myoglobin concentration in skeletal and cardiac muscle of bar-headed geese. *Comp. Biochem. Physiol. A* **100A,** 349–352.

Saunders, D. K., and Klemm, R. D. (1991). Seasonal changes in the skeletal, cardiac and smooth muscle of blue-winged teal. *Proc. Kans. Acad. Sci.* **10,** 29.

Scanes, C. G., Jallageas, M., and Assenmacher, I. (1980). Seasonal variation in the circulating concentrations of growth hormone on male Peking duck (*Anas platyrhynchos*) and teal (*Anas crecca*): Correlations with thyroidal function. *Gen. Comp. Endocrinol.* **41,** 76–79.

Schmidt-Nielsen, K., and Larimer, J. L. (1958). Oxygen dissociation curves of mammalian blood in relation to body size. *Am. J. Physiol.* **195,** 424–428.

Schmidt-Nielsen, K., Hainsworth, F. R., and Murrish, D. E. (1970). Counter-current heat exchange in the respiratory passages: Effect on water and heat balance. *Respir. Physiol.* **9,** 263–276.

Scholander, P. F. (1940). Experimental investigations on the respiratory function in diving mammals and birds. *Hvalradets Skr.* **22,** 1–131.

Scholander, P. F., and Irving, L. (1941). Experimental investigations on the respiration and diving of the Florida manatee. *J. Cell. Comp. Physiol.* **17,** 169–191.
Schuder, S., Wittenberg, J. B., Haseltine, B., and Wittenberg, B. A. (1979). Spectrophotometric determination of myoglobin in cardiac and skeletal muscle: Separation from hemoglobin by unit-exchange chromatography. *Anal. Biochem.* **92,** 473–481.
Shams, H., and Scheid, P. (1987). Respiration and blood gases in the duck exposed to normocapnic and hypercapnic hypoxia. *Respir. Physiol.* **67,** 1–12.
Shams, H., and Scheid, P. (1989). Efficiency of parabronchial gas exchange in deep hypoxia: Measurements in the resting duck. *Respir. Physiol.* **77,** 135–146.
Snyder, G. K., Byers, R. L., and Kayar, S. R. (1984). Effects of hypoxia on tissue capillarity in geese. *Respir. Physiol.* **58,** 151–160.
Staaden, R. (1984). The exercise physiology of the racing greyhound. Ph.D. Thesis, Murdoch University, Murdoch, Western Australia, Australia.
Stahl, W. R. (1967). Scaling of respiratory variables in mammals. *J. Appl. Physiol.* **22,** 453–460.
Stephenson, R., Turner, D. L., and Butler, P. J. (1989). The relationship between diving activity and oxygen storage capacity in the tufted duck (*Aythya fuligula*). *J. Exp. Biol.* **141,** 265–275.
Stoeckenius, M. (1963). Die Kapillarisierung verschiedener Volgelgehirne. *Gegenbaurs Morphol. Jahrb.* **105,** 343–364.
Suarez, R. K., Brown, G. S., and Hochachka, P. W. (1986). Metabolic sources of energy for hummingbird flight. *Am. J. Physiol.* **251,** R537–R542.
Suarez, R. K., Lighton, J. R. B., Moyes, C. D., Brown, G. S., Gass, C. L., and Hochachka, P. W. (1990). Fuel selection in rufous hummingbirds: Ecological implications of metabolic biochemistry. *Proc. Natl. Acad. Sci. U.S.A.* **87,** 9207–9210.
Suarez, R. K., Lighton, J. R. B., Brown, G. S., and Mathieu-Costello, O. (1991). Mitochondrial respiration in hummingbird flight muscles. *Proc. Natl. Acad. Sci. U.S.A.* **88,** 4870–4873.
Swan, L. W. (1961). The ecology of the high Himalayas. *Sci. Am.* **205,** 68–78.
Swan, L. W. (1970). Goose of the Himalayas. *Nat. Hist. (N.Y.)* **79,** 68–75.
Swatland, H. J. (1984). Intracellular distribution of succinate dehydrogenase activity in skeletal muscle fibers of geese. *Can. J. Zool.* **62,** 235–240.
Thomas, V. G. (1985). Myoglobin levels and mATPase activity in pectoral muscles of spruce and ruffed grouse (Aves: Tetraoninae). *Comp. Biochem. Physiol. A* **81A,** 181–184.
Tomlinson, J. T. (1963). Breathing of birds in flight. *Condor* **65,** 514–516.
Torre-Bueno, J. R. (1976). Temperature regulation and heat dissipation during flight in birds. *J. Exp. Biol.* **65,** 471–482.
Torre-Bueno, J. R. (1978a). Respiration during flight in birds. *In* "Respiratory Function in Birds, Adult and Embryonic" (J. Piiper, ed.), pp. 89–94. Springer-Verlag, Berlin.
Torre-Bueno, J. R. (1978b). Evaporative cooling and water balance during flight in birds. *J. Exp. Biol.* **75,** 231–236.
Torre-Bueno, J. R., and Larochelle, J. (1978). The metabolic cost of flight in unrestrained birds. *J. Exp. Biol.* **75,** 223–229.
Trivelpiece, W. Z., Bengston, J. L., Trivelpiece, S. G., and Volkman, N. J. (1986). Foraging behavior of gentoo and chinstrap penguins as determined by new radiotelemetry techniques. *Auk* **103,** 77–781.
Tucker, V. A. (1968a). Respiratory physiology of house sparrows in relation to high-altitude flight. *J. Exp. Biol.* **48,** 55–66.

Tucker, V. A. (1968b). Respiratory exchange and evaporative water loss in the flying budgerigar. *J. Exp. Biol.* **48,** 67–87.

Tucker, V. A. (1969). The energetics of bird flight. *Sci. Am.* **220,** 70–78.

Tucker, V. A. (1971). Flight energetics in birds. *Am. Zool.* **11,** 115–124.

Turek, Z., Ringnalda, B. E. M., Graudtner, M., and Kreuzer, R. (1973). Myoglobin distribution in the heart of growing rats exposed to a simulated altitude of 3500 m in their youth or born in the low pressure chamber. *Pflüegers Arch.* **340,** 1–10.

Vetterlein, F., Hemeling, H., Sammler, J., Petho A., and Schmidt, G. (1989). Hypoxia-induced acute changes in capillary and fiber density and capillary red cell distribution in the rat heart. *Circ. Res.* **64,** 742–752.

Walkinshaw, L. (1973). "Cranes of the World." Winchester, New York.

Ward, P., and Jones P. J. (1977). Pre-migratory fattening in three races of the red-billed quelea *Quelea quelea* (Aves: Ploceidae), an intra-tropical migrant. *J. Zool.* **181,** 43–56.

Weathers, W. W., and Snyder, G. D. (1974). Functional acclimation of Japanese quail to simulated altitude. *J. Comp. Physiol.* **93,** 127–137.

Weber, R. E., Hemmingsen, E. A., and Johansen, K. (1974). Functional and biochemical studies of penguin myoglobin. *Comp. Biochem. Physiol. B* **49B,** 197–214.

Weber, R. E., Hiebel, I., and Braunitzer, G. (1988). High altitude and hemoglobin functions in the vultures *Gyps ruppellii* and *Aegypius monachus*. *Biol. Chem. Hoppe-Seyler* **369,** 233–240.

Weibel, E. R. (1972). Morphometric estimation of pulmonary diffusion capacity. V. Comparative morphometry of alveolar lungs. *Respir. Physiol.* **14,** 26–43.

Weinstein, Y., Bernstein, M. H., Bickler, P. E., Gonzales, D. V., Samaniego, F. C., and Escobedo, M. A. (1985). Blood respiratory properties in pigeons at high altitudes: Effects of acclimation. *Am. J. Physiol.* **249,** R765–R775.

West, J. B. (1984). High living: Lessons from extreme altitude. *Am. Rev. Respir. Dis.* **130,** 917–923.

Wittenberg, B. A., and Wittenberg, J. B. (1987). Myoglobin-mediated oxygen delivery to mitochondria of isolated cardiac myocytes. *Proc. Natl. Acad. Sci. U.S.A.* **84,** 7503–7507.

Wittenberg, B. A., and Wittenberg, J. B. (1989). Transport of oxygen in muscle. *Annu. Rev. Physiol.* **51,** 857–878.

Young, D. A., and Boag, D. A. (1982). Changes in physical condition of male mallards (*Anas platyrhynchos*) during moult. *Can. J. Zool.* **60,** 3220–3226.

ADVANCES IN VETERINARY SCIENCE AND COMPARATIVE MEDICINE, VOL. 38B

Exercise Performance of Mammals: An Allometric Perspective

STAN L. LINDSTEDT AND RICKEY G. THOMAS

Department of Biological Sciences, Northern Arizona University, Flagstaff, Arizona 86011

I. Introduction
II. Body Size and Aerobic Exercise
  A. Body Size and Oxygen Uptake
  B. Body Size and Energy Cost of Transport
III. Muscle: The Source of Oxygen Demand
IV. Oxygen Delivery to Muscles
  A. Tissue Diffusing Capacity
  B. Cardiovascular Oxygen Delivery
V. Pulmonary Oxygen Uptake
  A. Pulmonary Ventilation
  B. Respiratory Mechanics
VI. Kinetics of Oxygen Transfer
VII. Body Size and Aerobic Performance: Emergent Properties
VIII. Summary
References

I. Introduction

In 1928 the English philosopher and physiologist J. B. S. Haldane began an engaging essay, "On Being the Right Size," with a discussion of the dimensions of the giants Pope and Pagan from *Pilgrim's Progress*. He calculated that if the giants were 10 times Christian's height and identical in shape, they would have been 10 × 10 × 10 or 1000 times his mass; however, Haldane concluded, the cross-sectional areas of their bones would have been only 100 times those of Christian. "As the human thigh bone breaks under about 10 times the human weight, Pope and Pagan would have broken their bones every time they took a step. This is doubtless why they were sitting down in the picture I remember. But it lessens one's respect for Christian and Jack the Giant Killer" (Haldane, 1956). Natural selection, unlike John Bunyan, the author of *Pilgrim's Progress*, has ensured that the limb bones of both shrews and bison have roughly the same safety factor for failure.

Among the mammals, the largest (blue whale) is 75,000,000 times the size of the smallest (the 2-g Etruscan shrew). What engineers have long known has only recently been studied with intensity by biologists: there must be trade-offs that accompany size changes. Scaling up a bridge or a mammal 100,000 times in size requires more than merely making larger parts. It is apparent that there are a suite of body-size-dependent physical laws that dictate many, if not most, features of mammalian design. Primary among these are some evident laws constraining the biomechanics, energetics, and performance of exercise in mammals. These laws become evident only when mammals of different size are directly compared. The focus of this paper will be to summarize many of those comparisons as they relate to exercise performance.

Athletic mammals can be defined as any that have achieved distinction by exploiting an extreme of muscle design and function. Power athletes like cheetahs, or human sprinters, fuel their short bursts of sprinting entirely via glycolysis; they have no need for increased oxygen supply to the muscles during exercise. Their performance is thus limited only by peak power output, hence the large mass, cross-sectional areas, and contraction velocities of their muscle fibers.

However, a different kind of athleticism is involved when sustained high power output is required over long periods. Energy liberation via anaerobic glycolysis of on-board fuel supply is insufficient to achieve prolonged high power output. Aerobic athletes must rely on the delivery of both fuel and oxygen in the right amounts at the right places and the simultaneous removal of heat and carbon dioxide. Thus, these athletes have the dual problem of having sufficient muscle mass to provide the necessary power, as well as the logistic problem of refueling and exhausting the muscles while they are actively working. We will consider only aerobic exercise in this paper, and thus the design and function of the mammalian respiratory system as a function of body mass among the mammals.

Historically, mammalian respiratory system function has been studied and is best understood in resting animals. However, as an animal begins to exercise, oxygen uptake increases dramatically. Further, it is only when the flow of oxygen through the respiratory system reaches its maximum that any of the structures through which it flows is potentially utilized to its full capacity. Thus, any constraints or rate-limiting steps in the cascade of structures responsible for the uptake, delivery and utilization of oxygen will be apparent only during maximum oxygen uptake ($\dot{V}_{O_2max}$).

We define $\dot{V}_{O_2max}$ as the plateau of oxygen uptake in an exercising animal at which oxygen consumption becomes uncoupled from its otherwise nearly linear increase that occurs with increasing power output. In other words, as the exercise intensity (power output) increases, so does $\dot{V}_{O_2}$, but only to a point. At work loads exceeding that critical power output, $\dot{V}_{O_2}$ remains virtually unchanged and the additional power is supplied by anaerobic glycolysis. It should be emphasized that this $\dot{V}_{O_2max}$ plateau is not a steady-state condition; animals can maintain $\dot{V}_{O_2max}$ for only short periods of time because of lactate accumulation. However, $\dot{V}_{O_2max}$ is a useful condition for interspecific comparisons for two reasons. First, $\dot{V}_{O_2max}$ represents an identical physiological state in which radically different animals can be directly compared to one another. Second, it is *the* upper limit to the flow rate of oxygen through animals' respiratory systems. For these reasons, much of the following discussion will focus on the causes and consequences of the body size dependency of $\dot{V}_{O_2max}$.

II. Body Size and Aerobic Exercise

The demand for oxygen in an exercising animal is set solely by the energetics of the working muscles; an animal's oxygen consumption can *never* exceed that demand. However, the muscles' demand for oxygen *may* exceed the capacity of the respiratory system to supply oxygen. To examine the direct effects of body size on aerobic exercise performance, we must first consider the design of those structures through which oxygen flows enroute to the mitochondria of the exercising muscles, that is, ventilation, pulmonary uptake, cardiovascular delivery, and tissue diffusion, as well as the muscular utilization of oxygen.

Ultimately, in aerobically exercising animals the requirement for ATP in the muscle fibers is translated into demand for oxygen to power the oxidative phosphorylation occurring in the mitochondria. Aerobic performance may be constrained by:

1. Delivery, that is, the ability of the pulmonary and cardiovascular systems to provide adequate fuel and oxygen to the periphery;
2. Tissue uptake, that is, the release of oxygen from the erythrocytes

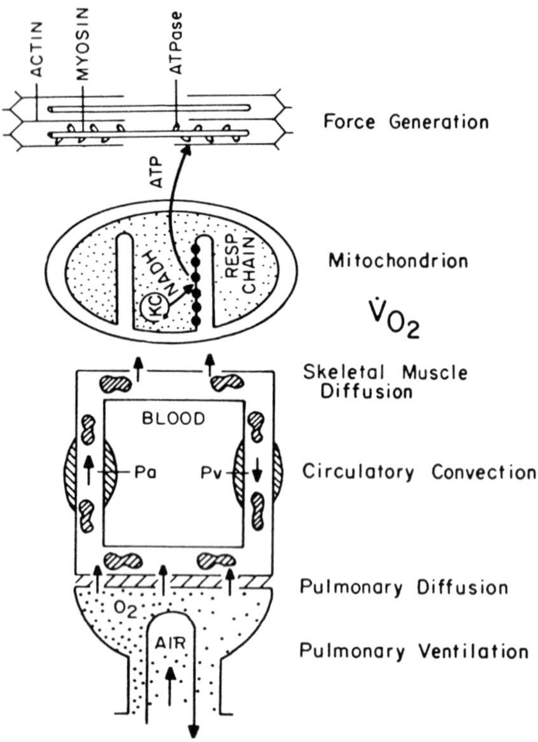

FIG. 1. This now familiar schematic of the mammalian respiratory system was first conceived by Dejours (1966) and later elaborated by Weibel and Taylor (1981). We have chosen to rotate the figure 180° because ultimately the demand for oxygen is determined not by the lung, but by the energetics of the working muscle. That demand is met only when sufficient oxygen is pumped by ventilation into the lung, moves by diffusion into the blood, is pumped to the muscle tissue by the heart, and finally moves by diffusion into the mitochondrial inner membranes, the site of oxidative phosphorylation. Collectively, function in those structures through which oxygen flows on its journey to the mitochondria must be subservient to the demand set by the working muscles.

and its diffusion into the muscles, including the facilitated diffusion of oxygen by myoglobin; and/or

3. The muscles' aerobic capacity, that is, the oxidative enzymes present in the muscle that are available to process the oxygen.

We choose the definition of the respiratory system to include all of those structures responsible for the pulmonary uptake, cardiovascular delivery, and tissue utilization of oxygen (Fig. 1). This paper will blend data with fanciful speculation to discuss the apparent constraints of body size on locomotion, muscle energetics, and the structure and function of each of these component parts of the mammalian respiratory system. How do each of these structures vary with body size? What is the relationship between the amount of structure and its use during exercise?

A. Body Size and Oxygen Uptake

The relationship between mammalian body size and weight-specific oxygen consumption has been the subject of considerable discussion for over a century (Rübner, 1883; Kleiber, 1932; Brody, 1945). This and other allometric relations are best described in the form of power law equations relating some variable (Y) to body mass (M_b) as $Y = aM_b^b$. When plotted on logarithmic coordinates, this expression describes a straight line with a slope of b. [Other representations, such as reduced major axis transformation or multiple regression, have been suggested to be statistically more correct, however, they lack the historical perspective and the apparent biological insight provided by the simpler and more direct power law equation, which will be used here (see Garland, 1983).] Though both the causes as well as the exact values of the relationship are still debated, basal metabolism has been convincingly and repeatedly demonstrated to vary as the $\frac{3}{4}$ power of body mass ($b = 0.75$). Because all animals at steady state must be in thermal equilibrium with their surroundings, this interspecific $\frac{3}{4}$ exponent must reveal the size-dependent rate of heat loss in resting, postabsorptive thermoneutral mammals. The scaling of maximum oxygen uptake in most (i.e., nonelite) exercising mammals has a slightly higher exponential relationship, $b = 0.85$ (Taylor *et al.*, 1981). However, exercising mammals must dissipate, not conserve heat. Because 95% of the oxygen flowing through the lungs of animals exercising at $\dot{V}_{O_2max}$ is destined for a single sink in the skeletal muscle mitochondria, any interspecific design principles implicated in setting the body-size-dependent scaling of $\dot{V}_{O_2max}$ are more likely to be biomechanical or energetic rather than thermal.

In addition to body size scaling of resting and maximum oxygen uptake, another conspicuous body-size-dependent pattern is the scaling of the cost of transport.

B. Body Size and Energy Cost of Transport

The cost of transport, defined as the oxygen cost to transport a unit mass a unit distance, can be calculated for any moving machine (Tucker, 1975). When steady-state weight-specific oxygen consumption is plotted against running velocity in a running mammal, the resultant relationship is represented by a nearly straight line (Schmidt-Nielsen, 1970). The slope of this relationship (volume of oxygen consumed per unit body mass per distance covered) is a measure of the energy cost of transport. This definition ignores the apparent cost for maintaining posture, the *Y*-intercept when the line is extrapolated to zero running velocity, which exceeds resting metabolism (Taylor, 1973). Thus defined, the energy cost of transport among the mammals scales as $M_b^{-0.30}$ (Taylor *et al.*, 1982). [If lizards, birds, and mammals are lumped together, the exponent changes little, to −0.28 (Fedak and Seeherman, 1979).] However, at a single, physiologically equivalent speed (the trot–gallop transition), the energy required to move a unit of mass a unit distance scales with a much lower exponent, $M_b^{-0.16}$ (Taylor *et al.*, 1982).

What constraints does body size place on the cost of transport, maximum oxygen uptake, respiratory system design, and the bioenergetics of skeletal muscle? Although the number of unresolved issues exceeds those for which we have substantial understanding, insights into each of the components of this question are surfacing. Because skeletal muscle is both the source of the demand for oxygen as well as the nearly exclusive sink for the oxygen flowing through the respiratory system at $\dot{V}_{O_2max}$, we will first consider the body size constraints in the design and energetics of skeletal muscle. We will then work back through the respiratory system, finishing with a discussion of diffusion and mechanics in the respiratory gas exchanger.

III. Muscle: The Source of Oxygen Demand

Across the class of mammals, the *structural* similarities of skeletal muscle are far more striking than are any differences (Table I). For example, all mammals have about 200 skeletal muscles that together comprise roughly 40% of the total body mass (Lindstedt *et al.*, 1993). Further, the skeletal muscles from all mammals are nearly identical both microscopically and macroscopically. Fiber cross-sectional area, sarcomere length, myofilament orientation, and myofibrillar dimensions are essentially invariant among the mammals (Eisenberg, 1983). Moreover, the noncontractile components—mitochondria, capillaries, and sarcoplasmic reticulum—are all structurally indistinguishable across mammalian body size. Finally, all mammalian skeletal muscles utilize identical biochemical pathways of energy liberation and, likely, even biochemical control mechanisms (Conley and Kushmerick, 1993).

TABLE I

VARIABLES THAT ARE INDEPENDENT OF BODY SIZE[a]

Variable	Units	a	b	N	r^2	Reference
Hemoglobin O_2 capacity	ml O_2 g^{-1} Hb	1.34	0.002	17	0.04	Larimer (1959)
[Hb]	g Hb dl^{-1}	12.9	0.011	16	0.29	Larimer (1959)
[Mb] muscle	g kg wet wt^{-1}	8.61	−0.023	5	0.19	Wittenberg (1970)
[Mb] heart	g kg wet wt^{-1}	3.23	0.024	7	0.14	Wittenberg (1970)
$Ca_{O_2}-C\bar{v}_{O_2}$	ml O_2 dl^{-1}	14.9	0	8	—	Taylor *et al.* (1988)
τ_h	μm	0.42	0.05	36	0.63	Gehr *et al.* (1981)
$\Theta(O_2)$	ml ml^{-1} $Torr^{-1}$ min^{-1}	1.21	0.06	6	0.11	Holland and Forster (1966)
Hct	%	39.8	−0.01	8	0.02	Altman and Dittmer (1961)
Arterial P_{O_2}	Torr	91.2	0.014	12	0.13	Lahiri (1975)
Arterial P_{CO_2}	Torr	37.2	0.018	9	0.29	Lahiri (1975)
P_{50}	Torr	34.6	−0.05	16	—	Schmidt-Nielsen and Larimer (1958)
$[HCO_3^-]$	m*M*	21.4	0.048	12	0.54	Lahiri (1975)
S_{im}	m^2 cm^{-3}	29.4	0.002	6	0.00	Hoppeler and Lindstedt (1985)

[a] For each variable, a is the intercept and b the exponent of the least-squares allometric regression, N is the number of data in the regression, and r^2 is the coefficient of determination. [Hb] is blood hemoglobin concentration, [Mb] is myoglobin concentration, $Ca_{O_2}-C\bar{v}_{O_2}$ is arterio-venous O_2 difference, τ_h is harmonic mean thickness of the alveolar membrane, Θ_{O_2} is the kinetic binding constant of O_2 with hemoglobin, P_{50} is half-saturation pressure of hemoglobin, and S_{im} is surface density of the inner mitochondrial membrane.

Thus, the basic blueprint of skeletal muscle design is essentially invariant among the mammals.

Similarly, *functional* similarities in skeletal muscle outweigh differences. The peak force per cross-sectional area of muscle is roughly constant, independent of the size of the animal whence it comes (about 50 N cm^{-2}; Goldspink, 1980; Rome *et al.*, 1990). Though there are significant differences in muscle histochemistry and biochemistry among different muscle fiber types, these differences are generally greater comparing different muscles *within* an individual animal than comparing the same muscles *among* mammals differing in body mass. All mammals seem to have an identical suite of muscle fiber types differing in characteristic ways in their metabolic and contractile properties (Ariano *et al.*, 1973). Further, the distribution of those fiber types is generally consistent across mammals of all sizes. Slow oxidative, Type I fibers tend to be assembled in the smallest motor units, are usually the deepest fibers near the bone, and are congregated in the motor units recruited first. In all mammals, these fibers have the smallest cross-sectional area and are responsible for aerobic tasks. In contrast, the fast glycolytic, Type IIB fibers are clustered in large motor units closer to the periphery of the muscle. In all mammals, these rapidly fatiguing fibers are the largest fibers in cross-sectional area and are located in the final motor units recruited, when maximum strength is required. Because the *soleus* muscles of a mouse and a moose are nearly exclusively Type I fibers, they have more in common with one another than a comparison of the *soleus* with the mixed-fiber *gastrocnemius* muscle within either species. It is now recognized that rather than two isozymes of myosin (slow versus fast), there are seven or eight isoforms of myosin among the mammals (D. Pette, personal communication), and these are likely shared isozymes common to most, if not all, mammals.

Despite these conspicuous similarities, the oxidative capacity of skeletal muscle varies unequivocally with body mass. In fact, body size alone is the single principle determinant of variation of muscle aerobic capacity among the mammals. Peak oxygen uptake of working muscle varies about 40-fold among the terrestrial mammals when viewed across the range of body sizes spanning over six orders of magnitude. During exercise, a gram of muscle from a shrew may consume as much oxygen in 2 s as does a gram of muscle from an elephant in 1 min. Thus, those structures coupled to the aerobic functioning of muscle vary with pronounced body mass dependency. The relative abundance, or *densities*, of capillaries and mitochondria and the *activities* of key oxidative enzymes are all much higher in small than large animals,

TABLE II

VARIABLES THAT SCALE ISOMETRICALLY WITH BODY SIZE[a]

Variable	Units	a	b	N	r^2	Reference
Lung volume	cm^3	66.1	0.99	32	0.94	Gehr *et al.* (1981)
Lung volume	cm^3	53.5	1.06	333	0.96	Stahl (1967)
Lung surface area	m^2	3.6	0.92	21	0.98	Gehr *et al.* (1981)
Lung capillary volume	ml	3.20	1.0	36	0.94	Gehr *et al.* (1981)
Muscle capillary volume	ml	5.88	0.94	19	—	Taylor *et al.* (1988)
$D_{L_{O_2}}$ (Physio)	ml O_2 s^{-1} $Torr^{-1}$	0.01	0.99	—	0.99	O'Neil and Leith (1980)
$D_{L_{O_2}}$ (Morpho)	ml O_2 s^{-1} $Torr^{-1}$	0.05	0.99	36	0.99	Gehr *et al.* (1981)
Dead space	cm^3	2.76	0.96	52	0.90	Stahl (1967)
Pulmonary compliance	ml cm^{-1} H_2O	1.56	1.04	114	0.98	Stahl (1967)
Lung surface area	m^2	3.34	0.95	36	0.99	Gehr *et al.* (1981)
Blood volume	ml	75.6	0.99	—	0.99	Prothero (1980)
Heart mass	kg	0.0058	0.98	200	0.98	Stahl (1967)
Stroke volume	ml	0.66	1.05	73	—	Holt *et al.* (1968)
Muscle mass	kg	0.385	1.02	4	0.99	Lindstedt *et al.* (1993)
Total hemoglobin	g	9.75	1.00	—	—	Prothero (1980) Larimer (1959)
Total muscle myoglobin	g	3.14	1.00	—	—	Wittenberg (1970) Lindstedt *et al.* (1993)

[a] Columns as in Table I. $D_{L_{O_2}}$ is pulmonary diffusing capacity for O_2 measured physiologically (Physio) or morphometrically (Morpho).

despite the apparent lack of discernible differences in their specific *structures* (Table II). Those differences in mean mitochondrial volume density and capillary length density are directly linked to the muscles' maximum aerobic capacities across mammalian, and probably avian, species. Hence, across a broad range of mammals running at $\dot{V}_{O_2max}$, the oxygen flow to the muscles per unit of capillary volume (0.3 ml O_2 ml^{-1} s^{-1}) and the maximum oxygen uptake per unit surface area of inner mitochondrial membrane (75,000 O_2 molecules μm^{-2} s^{-1}) are roughly constant (Table I; Lindstedt *et al.*, 1988; Wells, 1990; Hoppeler and Billeter, 1991; Conley and Kushmerick, 1993).

Despite notable exceptions (see the following and Jones and Lindstedt, 1993), the "default value" is that body size alone mandates changes in muscle oxidative energetics. This change in energetics is not merely a function of change in external power output. Taylor and colleagues (1980; Heglund *et al.*, 1982a) have demonstrated that external (weight-specific) work performed does not vary with body mass, however, the energy required to perform that work does scale allometrically with body mass (Heglund *et al.*, 1982b). What is the cause of this scaling of muscle aerobic energetics?

Body Size, Mechanics, and Energetics of Terrestrial Locomotion

Small mammals run with higher stride frequencies than do large mammals. Heglund *et al.* (1974) first reported that stride frequency varies regularly as $M_b^{-0.14}$ and stride length as $M_b^{0.38}$ when quadrupedal mammals running at the trot–gallop transition are compared. Following confirmation with little numerical change ($M_b^{-0.15}$; Heglund and Taylor, 1988), the scaling of stride frequency received strong theoretical support from Taylor (1985), who likened stride frequency scaling while running to a single mass and spring model, somewhat analogous to a resonating frequency. This same analogy was subsequently elaborated upon by McMahon *et al.* (1987), who characterized stride frequency as being set by a single, body-size-independent "Groucho number." Alexander (1989) presented a model based on analysis of Froude numbers among "dynamically similar" mammals, and likewise concluded that within any gait, stride frequency should vary approximately as $M_b^{-0.16}$. These and other studies support the concept that body mass alone is responsible for setting stride frequency in running mammals (Fig. 2). Once stride frequency is set, several aspects of muscle energetics may be similarly constrained via setting of the body-size-dependent intrinsic rate of muscle shortening.

The intrinsic rate of shortening (calculated as shortening velocity divided by muscle length, and functionally equivalent to sarcomere

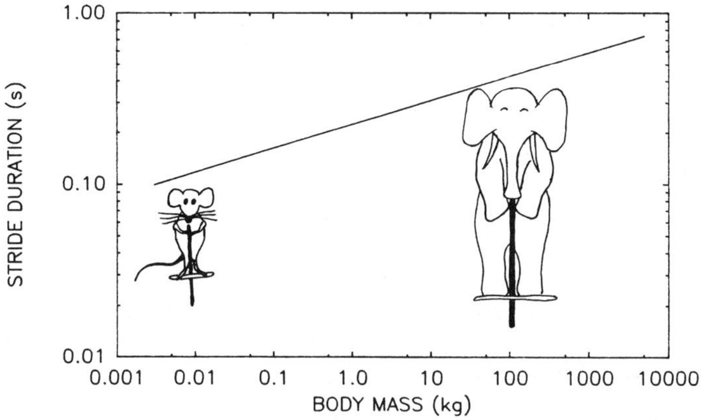

FIG. 2. Stride frequencies in running mammals are set in proportion to body mass. Larger mammals take longer strides less frequently. The predicted stride frequency of a 25-g mouse is over six times greater than that of a 5000-kg elephant (Heglund *et al.*, 1974; Taylor *et al.*, 1982).

shortening rate) is directly linked to rates of energy turnover in skeletal muscle (Bárány, 1967; Close, 1972). It now appears that the rate of cross-bridge cycling between actin and the myosin heads may ultimately set the energy cost of muscular contraction in direct proportion to the intrinsic rate of muscle shortening; there is a direct energy cost for increased speed (Goldspink, 1978; Crow and Kushmerick, 1982, 1983; Rall, 1985). Consequently, any size dependency in setting the intrinsic rate of muscle shortening may be reflected in the size-dependent energetics of exercising mammals. This concept has been theoretically and experimentally examined as a function of body mass among the mammals.

The biomechanics of running in mammals may set the rate of muscle shortening, which in turn sets energy demand. A. V. Hill (1950) was perhaps the first to suggest that the smallest mammals have the highest intrinsic rates of muscle shortening. *In vivo* muscle shortening has been estimated among running mammals by combining structural information (geometry of the knee and muscle lengths) with physiological data (knee excursion angles and muscle shortening times or stride frequencies). When these parameters are combined for running quadrupeds at the trot–gallop transition, this allometric analysis finds that the apparent intrinsic rate of muscle shortening (i.e., the change in muscle length divided by the time required for the change to occur divided by muscle length, units of s^{-1}) varies between $M_b^{-0.21}$ and

$M_b^{-0.23}$ for extensors and flexors of the knee, respectively (Lindstedt *et al.*, 1985).

The body size linking of the intrinsic rate of muscle shortening and energy cost of generating muscle force has been examined experimentally by Rome (1992), who predicted that the cost of generating force should be proportional to the maximum velocity of shortening (V_{max}) of the muscle fibers. He also concludes that V_{max} is strongly body size dependent, and that V_{max} is proportional to *in vivo* shortening velocities and inversely proportional to foot contact time in running mammals (hence, directly linked to stride frequency). Because animals of different body size appear to use similar muscle fiber recruitment patterns, scaling of energetic cost of force generation can be predicted by the V_{max} of the fiber types present in an animal. Measurements on single skinned muscle fibers have found that V_{max} in aerobic muscle fibers scales as $M_b^{-0.18}$ (Rome *et al.*, 1990), supporting the idea that muscle energetics must vary as a function of the obligate scaling of stride frequency and, therefore, intrinsic rate of muscle shortening. Small animals may be constrained to operate using muscles with high shortening velocities, resulting in a high size-dependent cost of generating force.

If the demand for energy is set as a dependent function of body size, then those structures in the muscle responsible for the supply and utilization of oxygen should also be quantitatively linked to intrinsic shortening rates. The body mass exponents describing the interspecific scaling of (a) oxygen uptake of the muscle (Weibel and Taylor, 1981); (b) volume density of mitochondria within the muscle (Taylor *et al.*, 1988); and (c) myosin ATPase activity in skeletal muscles across mammals in general (data from Syrovy and Gutmann, 1975, in Lindstedt *et al.*, 1985) are all apparently quantitatively coupled, thus their ratios are constant (Fig. 3). This skeletal muscle allometry suggests that there may be a causal link coupling muscle shortening rate to (a) the rate of demand for ATP production (mitochondrial volume); (b) the rate at which ATP may be hydrolyzed, providing the energy necessary to perform work (myosin ATPase activity); and finally (c) the maximum rate of oxygen uptake per unit muscle mass. It seems to us that body mass acts as the primary determinant of these variables. Thus, among the mammals there is an apparent size-dependent "default value," a starting point of sorts from which natural selection may operate to move certain species higher or lower. Calder (1984) suggests that comparisons of species to these allometric patterns provide evidence of "adaptive deviation."

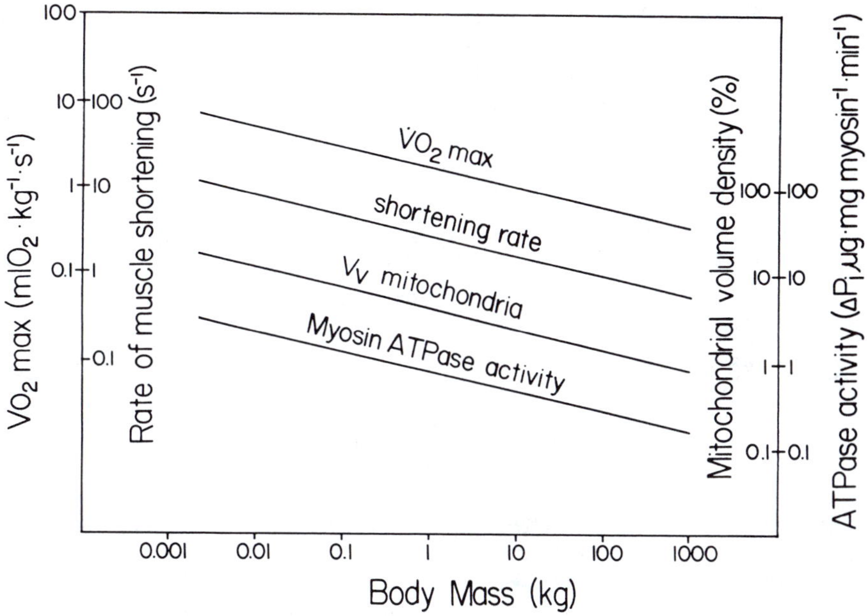

FIG. 3. Among the quadrupedal mammals there is an apparent linking of the intrinsic rate of muscle shortening of the knee extensors *in vivo* and their myosin ATPase activity (data from Syrovy and Gutmann, 1975), volume density of mitochondria (Lindstedt *et al.*, 1988), and weight-specific maximum oxygen uptake (Taylor *et al.*, 1981). Thus, despite each of these variables differing by about 25-fold between the largest and smallest terrestrial mammals, the *ratios* of these variables are apparently constant, independent of body size. From Lindstedt *et al.* (1985); reproduced with permission from the American Physiological Society.

IV. Oxygen Delivery to Muscles

A. Tissue Diffusing Capacity

A critical, that is, potentially rate limiting, diffusive step is the movement of oxygen from the erythrocytes to the sarcoplasm of the active muscle fiber (Wagner, 1988). Because this movement is by simple diffusion (not facilitated by a respiratory pigment) it requires a relatively large partial pressure driving force (Honig *et al.*, 1992). In contrast, the P_{O_2} within the sarcoplasm is thought to be roughly constant because of the presence of myoglobin. We can examine the structural components contributing to the "tissue diffusing capacity" for

PO 79
50
83
15
30

oxygen as a function of body size. First, and perhaps most critically, the total capillary volume density (capillary volume per unit volume of muscle) is nearly constant among the mammals (Taylor *et al.*, 1988), as is the capillary volume density in the lung (Table I). Second, because capillary diameter seems to be constant across body size, capillary surface area per unit volume of muscle is also roughly constant. Finally, the capillary P_{O_2} is likely invariant, however, because the half-saturation pressure for hemoglobin (P_{50}) is size dependent, small mammals may manage to have slightly higher oxygen gradients at the tissues than larger mammals. Taken together, these components of tissue diffusing capacity suggest that it is most likely a linear function of body size.

B. Cardiovascular Oxygen Delivery

Regardless of the magnitude of the demand for oxygen by the working muscles, their combined oxygen uptake can never exceed the rate at which oxygen is delivered by the cardiovascular system. The task of the cardiovascular system during exercise is to deliver oxygen, where needed, and in the correct amount. Perhaps for this reason, for over eight decades it has been suggested that cardiovascular delivery limits the muscles' aerobic capacity (e.g., Verzar, 1912; Hill and Lupton, 1923; Brooks and Fahey, 1984). Because both cardiac output and arterial oxygen concentration have well-defined upper limits, it is not unreasonable to expect that together these could function to set the upper limit to oxygen delivery and, therefore, $\dot{V}_{O_2max}$.

There is an abundant literature relating to tests of this hypothesis, perhaps in part because this is the simplest step in the oxygen transport cascade to manipulate experimentally. Thus, if the delivery capacity is reduced, for example, by anemia, carbon monoxide, or hypoxia, the result is a reduction in $\dot{V}_{O_2max}$ (Hillman, 1980; Longworth, 1993). Although results of these "reduced delivery" experiments have been presented as evidence supporting the delivery limitation hypothesis, these experiments do not support delivery limitation any more than delivery sufficiency (i.e., no excess capacity). However, experiments that have supplemented delivery (e.g., blood doping) are equivocal. For example, even when $\dot{V}_{O_2max}$ increases, the relative increase is considerably less than the changes in oxygen delivery (Lindstedt *et al.*, 1988). Though not now usually viewed as the single rate-limiting step, cardiovascular oxygen delivery is usually considered to be the most dominant single step in the limitation to $\dot{V}_{O_2max}$. Viewing the relation between delivery and oxygen uptake across mammals of varying body sizes provides no evidence in support of this view.

Whether or not $\dot{V}_{O_2max}$ is usually *limited* by oxygen availability, those factors that contribute to delivering oxygen to the tissues can be analyzed as functions of body size. Hemoglobin concentration, arterial oxygen concentration, and the kinetics of oxygen uptake in the lung and oxygen release in the tissues are all roughly constant and independent of body mass among the mammals (Table I). Once again, critical volumes, for example, stroke volume and total blood volume, are linear functions of body size (Table II). Because heart frequency scales, as do other frequencies, as $M_b^{-1/5}$–$M_b^{-1/4}$, both minimum and maximum cardiac output (the product of heart rate and stroke volume) and, therefore, oxygen delivery are strongly body size dependent, scaling as $M_b^{0.78}$ and $M_b^{0.81}$, respectively.

V. Pulmonary Oxygen Uptake

A. Pulmonary Ventilation

Pulmonary ventilation and pulmonary gas diffusion are the first two steps in the oxygen transport cascade. The cardiovascular system cannot deliver and the muscles cannot utilize oxygen unless it is first transported through the lungs into the blood. Hence, this step may also contribute to setting a limit to $\dot{V}_{O_2max}$, especially in endurance athletes or in hypoxia (Dempsey, 1986). Across all mammals, pulmonary volumes and capacities, like most others, are linear functions of body mass. Total lung volume, lung weight, pulmonary capillary volume, and alveolar surface area all scale linearly with body mass (Weibel *et al.*, 1981a; Table II). As a consequence, the capacity of the lung for oxygen diffusion (pulmonary diffusing capacity, $D_{L_{O_2}}$) is also a linear function of body mass, whether estimated physiologically or morphometrically (O'Neil and Leith, 1980; Weibel *et al.*, 1981a), despite the fact that both resting and maximum oxygen uptakes are strongly allometric, scaling between $M_b^{0.75}$ and $M_b^{0.85}$ (Table III). A 500-kg cow has, seemingly, 12 times more lung diffusing capacity per unit volume of oxygen consumed than does a 2-g shrew. This apparent "excess structure" has been presented as an allometric "paradox" of structure within the respiratory system (Weibel *et al.*, 1981b).

B. Respiratory Mechanics

Respiratory mechanics includes all displacements, shapes, and forces involved in breathing, that is, the flow resistive, elastic, and inertial elements of the respiratory system, and the pressures applied to it (Spells, 1969). Many of these have been examined comparatively (Agostoni *et al.*, 1959). From a mechanical viewpoint, the design features of the mammalian respiratory system may be constrained by two factors. The first of these is the need to satisfy the system's demand for gas exchange, requiring mechanisms for the diffusion between gas and blood that are tuned to the downstream demand for oxygen. This tuning necessitates appropriate mechanisms to ensure sufficient lung ventilation, as well as matching of ventilation with perfusion in the lung. The second factor is the need to satisfy any behavioral or environmental demands (e.g., posture or locomotion).

Comparative respiratory mechanics were first thoroughly examined as a function of mammalian body size by Leith (1976). He suggested a clustering of body mass exponents, with volumes and compliance scaling linearly with mass, and volume flows as $M_b^{3/4}$ (e.g., minute volume, oxygen consumption, and power of breathing; Table III). Hence, tidal volume is a linear function of body mass, as is the fraction of that volume that comprises the anatomical dead space (Table II). We could add that inspiratory pressures in the lung are likely invariant with (i.e., independent of) body size (Agostoni *et al.*, 1959). Leith (1982) concludes that tracheal dimensions are determined by mechanical considerations, namely, peak ventilatory flows and tracheal resistance to flow. Thus, the resistance to flow presented by the trachea is a constant fraction of the total resistance to flow in mammals of all body sizes.

Like other rates, respiratory rates vary as $M_b^{-0.26}$, at least at rest (Stahl, 1967). As a consequence, specific minute ventilation is 12 times greater in shrews than in cows, nearly the exact inverse of the scaling of $D\text{L}_{O_2}$ (Table III). Thus, the "paradox" of lung function and $D\text{L}_{O_2}$ may be resolved when we consider this critical rate function (Lindstedt, 1984). The net result of the size-dependent scaling of heart and respiratory rates is that small mammals maintain a greater A–a (alveolus to arterial blood) P_{O_2} gradient than large mammals, allowing for increased oxygen flux per unit alveolar membrane surface area (Lindstedt, 1984; Jones *et al.*, 1993; Longworth, 1993).

VI. Kinetics of Oxygen Transfer

The final allometric constraint that must affect the aerobic exercise performance of mammals is the kinetics of hemoglobin oxygen uptake

TABLE III

VARIABLES THAT SCALE ALLOMETRICALLY WITH BODY SIZE[a]

Variable	Units	*a*	*b*	*N*	r^2	Reference
Tracheal diameter	cm	0.45	0.39	45	—	Tenney and Bartlett (1967)
Tracheal length	cm	2.5	0.27	35	—	Tenney and Bartlett (1967)
Total airway resistance	cm H_2O s^{-1} $liter^{-1}$	24.4	−0.70	137	0.92	Stahl (1967)
Minute volume	ml s^{-1}	6.32	0.80	691	0.96	Stahl (1967)
Ventilatory rate	s^{-1}	0.89	−0.26	692	0.83	Stahl (1967)
Power of breathing	W	1.46	0.78	89	0.96	Stahl (1967)
Pulmonary RBC transit time	s	0.15	0.19	—	—	Gehr *et al.* (1981) Taylor *et al.* (1988)
Muscle RBC transit time	s	0.26	0.13	—	—	Taylor *et al.* (1988)
Subsarcolemmal mitochondria	%	21	−0.12	19	0.54	Lindstedt *et al.* (1993)
Mitochondrial volume	cm^3	30.2	0.87	19	—	Taylor *et al.* (1988)
Cytochrome oxidase	μl O_2 mg^{-1} h^{-1}	37	−0.22	4	0.9	Kunkel *et al.* (1956)
Carbonic anhydrase	EU · 100 μg Hb^{-1}	4.53	−0.10	13	0.77	Magid (1967)
Carbonic anhydrase	EU · 100 μg Hb^{-1}	3.82	−0.12	16	0.47	Larimer and Schmidt-Nielsen (1960)
Citrate synthase	μ*M* min^{-1} g^{-1} set	0.05	−0.12	20	0.37	Emmett and Hochachka (1981)
Lactate dehydrogenase	μ*M* min^{-1} g^{-1} wet	347	0.15	20	0.59	Emmett and Hochachka (1981)
$\dot{V}_{O_2max}$	ml s^{-1}	0.19	0.75	—	—	Kleiber (1932)
$\dot{V}_{O_2std}$	ml s^{-1}	1.67	0.85	45	0.98	Taylor *et al.* (1981)
Speed at $\dot{V}_{O_2max}$	m s^{-1}	2.34	0.12	22	0.45	Taylor *et al.* (1981)
In vivo Muscle Shortening Velocity at the Trot–Gallop Transition (T-GT)						
Knee extensors	s^{-1}	2.93	−0.23	—	—	Lindstedt *et al.* (1985)
Knee flexors	s^{-1}	—	−0.21	—	—	Lindstedt *et al.* (1985)

(continued)

TABLE III (*Continued*)

Variable	Units	a	b	N	r^2	Reference
Stride duration at T-GT	s	0.22	0.14	6	0.98	Heglund *et al.* (1974)
Stride length at T-GT	m	0.35	0.38	6	0.94	Heglund *et al.* (1974)
Cost of transport	ml O_2 m^{-1} kg^{-1}	3.59	−0.28	69	0.84	Fedak and Seeherman (1979)
Cost of transport	ml O_2 m^{-1} kg^{-1}	0.53	−0.32	79.0	0.95	Taylor *et al.* (1982)
V_{max} (aerobic fibers)	s^{-1}	0.99	−0.18	3	—	Rome *et al.* (1990)
Heart rate (rest)	s^{-1}	4.02	−0.25	447	0.77	Stahl (1967)
Heart rate (max)	s^{-1}	6.50	−0.15	14	0.86	Taylor *et al.* (1988) Baudinette (1978)
Cardiac output (rest)	ml s^{-1}	3.51	0.78	4	0.99	Lindstedt *et al.* (1993)
Cardiac output (max)	ml s^{-1}	21.9	0.81	8	—	Taylor *et al.* (1988)
Body surface area	m^2	0.11	0.65	100	0.90	Stahl (1967)

[a] Columns as in Table I.

and release. Again, we start with a listing of those variables that are constant, independent of body mass. There is no evidence that either alveolar or tissue P_{O_2} varies with body size, thus, the overall diffusing gradient for oxygen, lung to tissue, is invariant among the mammals (Fig. 1). Arterial P_{O_2} and P_{CO_2} are constant, as are concentrations of hemoglobin and myoglobin, and specific oxygen conductance (Θ_{O_2}) (Holland and Forster, 1966). Thus, the P_{O_2} driving forces at the lung and tissues, as well as the properties of the hemoglobin and myoglobin relating to the kinetics of oxygen transfer, are nearly invariant among the mammals.

Erythrocyte transit times through the pulmonary and muscle capillaries are linked in a 2:1 ratio, though transit times in small animals are much less than those in larger animals. The P_{O_2} of the mixed venous blood that enters the lung during exercise may be below 20 Torr. As it passes through the pulmonary capillaries the P_{O_2} increases, and under ideal conditions, alveolar and arterial P_{O_2} are at equilibrium when the blood leaves the lung. However, these conclusions, based on experimentation on large animals, especially humans, typically fail to consider the body size scaling of transit times; the transit times of erythrocytes through the pulmonary and muscle tissue capillaries are strongly body size dependent (Table III). Because the total capillary volume of all skeletal muscle tissue is about double that of the lungs, at $\dot{V}_{O_2max}$ the time available for oxygen unloading in the tissues will be no more than double that available for loading in the lungs. The actual time is certainly somewhat less than twice as much, because not all the muscle capillaries will be perfused. At $\dot{V}_{O_2max}$ in running quadrupeds, the majority of the aerobic fibers, and probably most of the muscle capillaries, are perfused. However, if only half are perfused, then the time for unloading in the muscle will be equal to that for loading in the lung. A "standard" 1-kg mammal exercising at $\dot{V}_{O_2max}$ will have about O.15 s to load oxygen in the lungs, and a maximum of 0.27 s to unload oxygen in the muscles. In a 10-g mouse, those times are reduced to roughly 0.06 s and a maximum of 0.14 s, respectively. The relative capillary volumes of the lungs and muscles and the resultant transit times are consistent with the observed kinetics of oxygen loading and unloading. Heidelberger and Reeves (1982) have shown with an *in vitro* preparation that oxygen loading requires about half as much time as oxygen unloading. Small animals have the least time available for oxygen diffusion in the lungs and tissues. This may help explain why muscle capillary density increases slightly in the smallest mammals and why P_{50} shifts to the right, both of which would facilitate unloading of oxygen. Because unloading requires twice as much time as loading, it may be unloading of oxygen that is the critical time in the smallest mammals.

VII. Body Size and Aerobic Performance: Emergent Properties

Thus far in this paper, we have reviewed and summarized size-dependent principles that govern the energetics and metabolism of aerobic exercise. These empirical equations can be used to generate predictive relationships by forming ratios of equations in a manner formalized by Stahl (1962, 1967; Calder, 1984). For example, by dividing respiratory and heart rates, Stahl demonstrated that all mammals have about 4.5 heart beats per breath at rest. Because the residual mass exponent is small (−0.01), this ratio is essentially independent of body mass (Stahl, 1967). By similar algebraic manipulation, we have calculated several other nearly body-size-independent ratios in Table IV. For example, all mammals have roughly twofold more (1.8×) muscle capillary volume than pulmonary capillary volume, 200 ml of blood per kilogram of muscle tissue, and 3.1 times as much hemoglobin as myoglobin. Additionally, total muscle mass is roughly 66 times the mass of the heart, and at rest, respiratory minute volume exceeds cardiac output by 1.8-fold. We also see constant proportionalities between mitochondrial volume density and the V_{max} of aerobic fibers, maximum oxygen consumption and the total area of inner mitochondrial membranes, and maximum oxygen uptake and the volume of muscle capillaries. These constant ratios may imply the existence of design constraints that transcend differences in body mass.

The body-size-dependent patterns that we have presented seem to fall into qualitative clusters. It must be emphasized that the following general principles describe the "default" traits common to most mammals, but by no means obligately found in all mammals. We view these size-dependent constraints as the departure point from which natural selection may operate, as well setting the upper/lower limits possible at any given size. Once these allometric patterns are identified, then those species that do not fit the pattern can be identified. Having presented that caveat, the identifiable clusters are:

1. In general, concentrations, tensions, and pressures are roughly *independent* of body size. Inspiratory and expiratory ventilation pressures, blood pressure and the partial pressures of O_2 and CO_2 in lungs, blood, and tissues do not apparently vary with body size. Likewise, concentrations of hemoglobin, myoglobin, and hematocrit

TABLE IV

EMERGENT PATTERNS FROM ALLOMETRIC ANALYSIS OF THE MAMMALIAN OXYGEN TRANSPORT SYSTEM

Dimensionless ratio	a	Residual mass exponent
Heart rate/respiratory rate	4.5	−0.01
Muscle mass/heart mass	66.3	0.04
Muscle capillary volume/lung capillary volume	1.84	−0.06
Total hemoglobin content/total myoglobin content	3.10	0.00
Arterial P_{O_2}/arterial P_{CO_2}	2.45	0.00
Total blood volume/stroke volume	115	−0.07
Minute volume/cardiac output (rest)	1.80	0.02
Dead space/lung volume	4.2	0.02

Ratio	Units	a	Residual mass exponent
Blood volume/muscle mass	ml kg^{-1}	197	−0.03
Mitochondrial volume kg^{-1}/V_{max} (Type I fibers)	cm^3 kg^{-1} s^{-1}	79	0.03
Cardiac output (max)/$\dot{V}_{O_2 max}$	ml O_2 ml^{-1} blood	0.076	0.04
$\dot{V}_{O_2 max}$/mitochondrial volume	ml O_2 s^{-1} ml^{-1}	0.055	−0.02
$\dot{V}_{O_2 max}$/mitochondrial inner membrane area	O_2 molecules μm^{-2} s^{-1}	75,000	−0.02
$\dot{V}_{O_2 max}$/muscle capillary volume	ml O_2 s^{-1} cm^{-3}	0.28	−0.08

are all roughly independent of body size. Red blood cell size does vary among the mammals, however, it does not do so systematically with body size. As a consequence, the diameter of capillaries in muscle tissue is also body mass independent.

2. Most volumes and capacities scale *linearly* with size (i.e., as a constant function of body mass). Mass of the heart is roughly constant among the mammals (0.5% of total body mass), as is the total blood volume (7.0%). As a result, stroke volume is a constant fraction of body mass, as is the fraction of the total blood volume pumped per beat, hence the number of heart beats required to circulate the entire volume of blood once is constant (about 115).
3. Virtually all time-dependent biological processes proceed more rapidly in smaller than in larger animals; biological frequencies seem to be set to a common size-dependent clock (Lindstedt and Calder,

1981). Most time-dependent variables related to muscle energetics and oxygen delivery and utilization scale *allometrically* with body mass; they are higher in small animals than in large ones. In general, biological rates, such as heart rate (both resting and maximal), vary between the $-\frac{1}{5}$ to $-\frac{1}{4}$ power of body mass. (Because frequency or rate is the reciprocal of time, biological times scale as their reciprocals, $\frac{1}{5}$ to $\frac{1}{4}$.) Among other rates scaling similarly are the intrinsic rate of muscle shortening and the transit times of blood through the lungs and muscles.

4. Finally, volume-rates are the product of a volume and a rate; these generally vary as $M_b^{1.0} \times M_b^{-0.20 \text{ to } -0.25} = M_b^{0.75 \text{ to } 0.80}$. For example, cardiac output is the product of a volume (stroke volume) and a frequency (heart rate); it varies, as do most volume-rates, as mass raised to the 0.78 and 0.81 power at rest and $\dot{V}_{O_2max}$, respectively. Oxygen consumption, minute volume, and power output all conform to this scaling rule.

Though the vast majority of mammals fall on the scaling lines described in this paper, several notable exceptions do not. Just as important in asking why most mammals fit the size-dependent pattern is the question: Why do many mammals NOT fit the pattern? Some premiere aerobic athletes, for example, canids, especially greyhounds, horses, and pronghorn antelope are the products of intense selection for increased aerobic capacity. Even though that selection can be either artificial (man-made) or natural, $\dot{V}_{O_2max}$ and, presumably, muscle power output in these species exceed that predicted allometrically by three- to fivefold. However, in all of these species there is a general coupling of oxygen uptake and those structures (lung diffusing capacity, cardiovascular oxygen delivery, and mitochondrial volume) responsible to support their aerobic performances. In general, these species conform to body size predictions for frequencies (e.g., stride frequency, heart and respiratory rates), but volumes (e.g., trachea, heart, and lungs) are much larger than predicted allometrically, in proportion to demand (Weibel and Taylor, 1981; Taylor *et al.*, 1987; Lindstedt *et al.*, 1991; Jones and Lindstedt, 1993). These animal athletes are the subject of several reviews (Jones *et al.*, 1989, 1993; Jones and Lindstedt, 1993).

Across the range of mammalian body size, it is apparent that body size alone is the greatest single determinant of aerobic exercise performance. Size alone seems to set the "default values" of structures and functions. The predicted range of weight-specific (maximum) aerobic capacity, comparing a 2-g shrew with a 500-kg elephant, is about 20-fold. We have presented several allometric equations that predict how several variables relating to oxygen uptake, transfer, and utilization scale with body mass.

VIII. Summary

We have examined aerobic exercise performance among the mammals with particular attention to the constraints that body size places on all aspects of muscle biomechanics, aerobic energetics, tissue oxygen diffusion, cardiovascular oxygen delivery, and pulmonary oxygen uptake. Several body-size-dependent patterns emerge that seemingly govern aerobic performance in mammals, with the caveat that at any given body size there is a range of aerobic capacities, the result of natural selection operating on the size-dependent "default values" of structure and function. Among these default values, the following apparent functional clusters surface:

1. In general, concentrations and pressures (e.g., of proteins and gases) are roughly *independent* of body size. Inspiratory and expiratory ventilation pressures, blood pressure and the partial pressures of O_2 and CO_2 in lungs, blood, and tissues do not vary with body size. Likewise, concentrations of hemoglobin, myoglobin, and hematocrit are independent of body size.
2. Most volumes and capacities scale *linearly* with body size (i.e., as a constant function of body mass). In addition to heart, lung, and total blood volumes, important examples relevant to exercise performance are the diffusing capacities for oxygen in the lung and, apparently, in the tissues.
3. Finally, most time-dependent variables related to oxygen delivery scale *allometrically* with body mass; they are of shorter duration in small animals than in large ones. Biological rates, for example, V_{max} of working muscle, heart and respiratory rates, and transit times of blood through the muscles and lungs, all vary roughly as the $-\frac{1}{5}$ to $-\frac{1}{4}$ power of body mass.

References

Agostoni, E., Thimm, F. F., and Fenn, W. O. (1959). Comparative features of the mechanics of breathing. *J. Appl. Physiol.* **14,** 679–683.

Alexander, R. (1989). Optimization and gaits in the locomotion of vertebrates. *Physiol. Rev.* **69,** 1199–1227.

Altman, P. L., and Dittmer, D. S. (eds.) (1961). "Biological Handbook." Fed. Am. Soc. Exp. Biol., Bethesda, MD.

Ariano, M. A., Armstrong, R. B., and Edgerton, V. R. (1973). Hindlimb muscle fiber population of five mammals. *J. Histochem. Cytochem.* **21,** 51–55.

Bárány, M. A. (1967). ATPase activity of myosin correlated with speed of muscle shortening. *J. Gen. Physiol.* **50,** 197–218.

Baudinette, R. V. (1978). Scaling of heart rate during locomotion in mammals. *J. Comp. Physiol.* **127,** 337–342.

Brody, S. (1945). "Bioenergetics and Growth." Reinhold, New York.

Brooks, G. A., and Fahey, T. D. (1984). "Exercise Physiology." Wiley, New York.

Calder, W. A., III (1984) "Size, Function and Life History." Harvard Univ. Press, Cambridge, MA.

Close, R. I. (1972). Dynamic properties of mammalian skeletal muscles. *Physiol. Rev.* **52,** 129–197.

Conley, K. E., and Kushmerick, M. J. (1993). Unifying principles in aerobic muscle energetics. *In* "The Vertebrate Gas Transport Cascade: Adaptations to Environment and Mode of Life" (J. E. P. W. Bicudo, ed.), pp. 304–313. CRC Press, Boca Raton, FL.

Crow, M. T., and Kushmerick, M. J. (1982). Chemical energetics of slow- and fast-twitch muscles of the mouse. *J. Gen. Physiol.* **79,** 147–166.

Crow, M. T., and Kushmerick, M. J. (1983). Correlated reduction of velocity of shortening and the rate of energy utilization in mouse fast-twitch muscle during a continuous tetanus. *J. Gen. Physiol.* **82,** 703–720.

Dejours, P. (1966). "Respiration." Oxford Univ. Press, New York.

Dempsey, J. A. (1986). Is the lung built for exercise? *Med. Sci. Sports Exercise* **18,** 143–155.

Eisenberg, B. R. (1983). Quantitative ultrastructure of mammalian skeletal muscle. *In* "Handbook of Physiology" (L. D. Peachey, R. H. Adrian, and S. Geiger, eds.), Sect. 10, pp. 73–112. Williams & Wilkins, Co., Baltimore, MD.

Emmett, B., and Hochachka, P. W. (1981). Scaling of oxidative and glycolytic enzymes in mammals. *Respir. Physiol.* **45,** 261–272.

Fedak, M. A., and Seeherman, H. J. (1979). Reappraisal of energetics of locomotion shows identical cost in bipeds and quadrupeds including ostrich and horse. *Nature (London)* **282,** 713–716.

Garland, T., Jr. (1983). The relation between maximal running speed and body mass in terrestrial mammals. *J. Zool.* **199,** 157–170.

Gehr, P., Mwangi, D. K., Amman, A., Maloiy, G. M. O., Taylor, C. R., and Weibel, E. R. (1981). Design of the mammalian respiratory system. V. Scaling morphometric pulmonary diffusing capacity to body mass: Wild and domestic animals. *Respir. Physiol.* **44,** 61–86.

Goldspink, G. (1978). Energy turnover during contraction of different types of muscle. *Int. Ser. Biomech.* **2,** 27–39.

Goldspink, G. (1980). Locomotion and the sliding filament mechanism. *In* "Aspects of Animal Movement" (H. Y. Elder and E. R. Trueman, eds.), pp. 1–26. Cambridge Univ. Press, Boston.

Haldane, J. B. S. (1956). On being the right size. *In* "The World of Mathematics II" (J. R. Newman, ed.), pp. 952–957. Simon & Schuster, New York.

Heglund, N. C., and Taylor, C. R. (1988). Speed, stride frequency and energy cost per stride: How do they change with body size and gait? *J. Exp. Biol.* **138,** 301–318.

Heglund, N. C., Taylor, C. R., and McMahon, T. A. (1974). Scaling stride frequency and gait to animal size: Mice to horses. *Science* **186,** 1112–1113.

Heglund, N. C., Cavagna, G. A., and Taylor, C. R. (1982a). Energetics and mechanics of terrestrial locomotion. III. Energy changes of the centre of mass as a function of speed and body size in birds and mammals. *J. Exp. Biol.* **79,** 41–56.

Heglund, N. C., Fedak, M. A., Taylor, C. R., and Cavagna, G. A. (1982b). Energetics and mechanics of terrestrial locomotion. IV. Total mechanical energy changes as a function of speed and body size in birds and mammals. *J. Exp. Biol.* **79,** 57–66.

Heidelberger, E., and Reeves, R. B. (1982). Factors affecting whole blood O_2 transfer kinetics: Implications for $\Theta(O_2)$. *J. Appl. Physiol.* **68,** 1865–1874.

Hill, A. V. (1950). The dimensions of animals and their muscular dynamics. *Sci. Prog. (Oxford)* **38,** 209–230.

Hill, A. V., and Lupton, H. (1923). Muscular exercise, lactic acid and the supply and utilization of oxygen. *Q. J. Med.* **16,** 135–171.

Hillman, S. S. (1980). The effect of anemia on metabolic performance in the frog, *Rana pipiens*. *J. Exp. Zool.* **211,** 107–111.

Holland, R. A. B., and Forster, R. E. (1966). The effect of size of red cells on the kinetics of their oxygen uptake. *J. Gen. Physiol.* **49,** 199–220.

Holt, J. P., Rhode, E. A., and Kines, H. (1968). Ventricular volumes and body weight in mammals. *Am. J. Physiol.* **215,** 704–715.

Honig, C. R., Connett, R. J., and Gayeski, T. E. J. (1992). O_2 transport and its interaction with metabolism: A systems view of aerobic capacity. *Med. Sci. Sports Exercise* **24,** 47–53.

Hoppeler, H., and Billeter, R. (1991). Conditions for oxygen and substrate transport in muscles in exercising mammals. *J. Exp. Biol.* **160,** 263–283.

Hoppeler, H., and Lindstedt, S. L. (1985). Malleability of skeletal muscle tissue in overcoming limitations: Structural elements. *J. Exp. Biol.* **115,** 355–364.

Jones, J. H., and Lindstedt, S. L. (1993). Limits to maximal performance. *Annu. Rev. Physiol.* **53,** 547–570.

Jones, J. H., Longworth, K. E., Lindholm, A., Conley, K. E., Karas, R. H., Kayar, S. K., and Taylor, C. R. (1989). Oxygen transport during exercise in large mammals. I. Adaptive variation in oxygen demand. *J. Appl. Physiol.* **67,** 862–870.

Jones, J. H., Birks, E. K., and Pascoe, J. R. (1993). Factors limiting aerobic performance. *In* "The Vertebrate Gas Transport Cascade: Adaptations to Environment and Mode of Life" (J. E. P. W. Bicudo, ed.), pp. 169–178. CRC Press, Boca Raton, FL.

Kleiber, M. (1932). Body size and metabolism. *Hilgardia* **6,** 315–353.

Kunkel, H. O., Spalding, J. F., De Franciscis, G., and Futrell, M. F. (1956). Cytochrome oxidase activity and body weight in rats and in three species of large animals. *Am. J. Physiol.* **186,** 203–206.

Lahiri, S. (1975). Blood oxygen affinity and alveolar ventilation in relation to body weight in mammals. *Am. J. Physiol.* **229,** 529–536.

Larimer, J. L. (1959). Hemoglobin concentration and oxygen capacity of mammalian blood. *J. Elisha Mitchell Sci. Soc.* **75,** 174–177.

Larimer, J. L., and Schmidt-Nielsen, K. (1960). A comparison of blood carbonic anhydrase of various mammals. *Comp. Biochem. Physiol.* **1,** 19–23.

Leith, D. E. (1976). Comparative mammalian respiratory mechanics. *Physiologist* **19,** 485–510.

Leith, D. E. (1982). Mammalian tracheal dimensions: Scaling and physiology. *J. Appl. Physiol.* **55,** 196–200.

Lindstedt, S. L. (1984). Pulmonary transit time and diffusing capacity in mammals. *Am. J. Physiol.* **246,** R384–R388.

Lindstedt, S. L., and Calder, W. A. (1981). Body size, physiological time and longevity of homeothermic animals. *Q. Rev. Biol.* **56,** 1–15.

Lindstedt, S. L., Hoppeler, H., Bard, K. M., and Thronson, H. A. (1985). Estimate of muscle shortening rate during locomotion. *Am. J. Physiol.* **249,** R699–R703.

Lindstedt, S. L., Wells, D. J., Jones, J. H., Hoppeler, H., and Thronson, H. J. (1988).

Limitations to aerobic performance in mammals: Interaction of structure and demand. *Int. J. Sports Med.* **3,** 210–217.
Lindstedt, S. L., Hokanson, J. F., Wells, D. J., Swain, S. D., Hoppeler, H., and Navarro, V. (1991). Running energetics in the pronghorn antelope. *Nature (London)* **353,** 748–750.
Lindstedt, S. L., Thomas, R. J., Nichols, S. D., and Robinson, D. (1993). Body size scaling of physiological variables among mammals. Submitted for publication.
Longworth, K. E. (1993). Scaling of pulmonary diffusing capacity. *In* "The Vertebrate Gas Transport Cascade: Adaptations to Environment and Mode of Life" (J. E. P. W. Bicudo, ed.), pp. 162–168. CRC Press, Boca Raton, FL.
Magid, E. (1967). Activity of carbonic anhydrase in mammalian blood in relation to body size. *Comp. Biochem. Physiol.* **21,** 357–360.
McMahon, T. A., Valiant, G., and Frederick, E. C. (1987). Groucho running. *J. Appl. Physiol.* **62,** 2326–2337.
O'Neil, J. J., and Leith, D. E. (1980). Lung diffusing capacity scaled in mammals from 25 g to 500 kg. *Fed. Proc., Fed. Am. Soc. Exp. Biol.* **39,** 972 (abstr.).
Prothero, J. W. (1980). Scaling of blood parameters in mammals. *Comp. Biochem. Physiol.* **67,** 649–657.
Rall, J. A. (1985). Energetic aspects of skeletal muscle contraction: Implications of fiber types. *Exercise Sport Sci. Rev.* **13,** 33–74.
Rome, L. C. (1992). Scaling of muscle fibres and locomotion. *J. Exp. Biol.* **168,** 243–252.
Rome, L. C., Sosnicki, A. A., and Goble, D. O. (1990). Maximum velocity of shortening of three fibre types from horse soleus muscle: Implications for scaling with body size. *J. Physiol. (London)* **431,** 173–185.
Rübner, M. (1883). Ueber den Einfluss der Korpergrosse auf Stoff- und Kraftwechsel. *Z. Biol. (Munich)* **19,** 535–562.
Schmidt-Nielsen, K. (1970). Energy metabolism, body size and problems of scaling. *Fed. Proc., Fed. Am. Soc. Exp. Biol.* **29,** 1524–1532.
Schmidt-Nielsen, K., and Larimer, J. L. (1958). Oxygen dissociation curves of mammalian blood in relation to body size. *Am. J. Physiol.* **195,** 424–428.
Spells, K. E. (1969). Comparative studies in lung mechanics based on a survey of literature data. *Respir. Physiol.* **8,** 37–57.
Stahl, W. R. (1962). Similarity and dimensional methods in biology. *Science* **37,** 205–209.
Stahl, W. R. (1967). Scaling of respiratory variables in mammals. *J. Appl. Physiol.* **22,** 453–460.
Syrovy, I., and Gutmann, E. (1975). Myosin from fast and slow skeletal muscle and cardiac muscles of mammals of different size. *Physiol. Bohemoslov.* **24,** 325–334.
Taylor, C. R. (1973). Energy cost of animal locomotion. *In* "Comparative Physiology" (L. Bolis, K. Schmidt-Nielsen, and S. H. P. Maidrell, eds.), pp. 23–42. North-Holland Publ., Amsterdam.
Taylor, C. R. (1985). Force development during sustained locomotion: A determinant of gait, speed and metabolic power. *J. Exp. Biol.* **115,** 253–262.
Taylor, C. R., Heglund, N. C., McMahon, T. A., and Looney, T. R. (1980). Energetic cost of generating muscular force during running. *J. Exp. Biol.* **86,** 9–18.
Taylor, C. R., Maloiy, G. M. O., Weibel, E. R., Langman, V. A., Kamau, J. M. Z., Seeherman, M. J., and Heglund, N. C. (1981). Design of the mammalian respiratory system. III. Scaling maximum aerobic capacity to body mass: Wild and domestic mammals. *Respir. Physiol.* **44,** 25–37.
Taylor, C. R., Heglund, N. C., and Maloiy, G. M. O. (1982). Energetics and mechanics of terrestrial locomotion. I. Metabolic energy consumption as a function of speed and body size in birds and mammals. *J. Exp. Biol.* **97,** 1–21.

Taylor, C. R., Karas, R. H., Weibel, E. R., and Hopeler, H. (1987). Adaptive variation in the mammalian respiratory system in relation to energetic demand. *Respir. Physiol.* **69,** 1–127.

Taylor, C. R., Longworth, K. E., and Hoppeler, H. (1988). Matching O_2 delivery to O_2 demand in muscle. II. Allometric variation in energy demand. *In* "Oxygen Transfer from Atmosphere to Tissues" (N. C. Gonzalez and M. R. Fedde, eds.), pp. 171–181. Plenum, New York and London.

Tenney, S. M., and Bartlett, D. (1967). Comparative quantitative morphology of the mammalian lung: Trachea. *Respir. Physiol.* **3,** 130–135.

Tucker, V. A. (1975). The energetic cost of moving about. *Am. Sci.* **63,** 413–419.

Verzar, F. (1912). The gaseous metabolism of striated muscle in warm-blooded animals. *J. Physiol. (London)* **44,** 243–258.

Wagner, P. D. (1988). Evidence for peripheral tissue diffusion limitation of maximal oxygen uptake. *In* "Oxygen Transfer from Atmosphere to Tissues" (N. C. Gonzalez and M. R. Fedde, eds.). Plenum, New York and London.

Weibel, E. R., and Taylor, C. R. (1981). Design of the mammalian respiratory system. *Respir. Physiol.* **44,** 1–164.

Weibel, E. R., Gehr, P., Cruz-Orive, L. M., Müller, A. E., Mwangi, D. K., and Haussener, V. (1981a). Design of the mammalian respiratory system. IV. Morphometric estimation of pulmonary diffusing capacity: Critical evaluation of a new sampling method. *Respir. Physiol.* **44,** 39–60.

Weibel, E. R., Taylor, C. R., Gehr, P., Hoppeler, H., Mathieu, O., and Maloiy, G. M. O. (1981b). Design of the mammalian respiratory system. IX. Functional and structural limits for oxygen flow. *Respir. Physiol.* **44,** 151–164.

Wells, D. J. (1990). Hummingbird flight physiology: Muscle performance and ecological constraints. Ph.D. Dissertation, University of Wyoming, Laramie.

Wittenberg, J. (1970). Myoglobin-facilitated oxygen diffusion: Role of myoglobin in oxygen entry into muscle. *Physiol. Rev.* **50,** 559–636.

ADVANCES IN VETERINARY SCIENCE AND COMPARATIVE MEDICINE, VOL. 38B

Beyond the Vertebrates: Achieving Maximum Power during Flight in Insects and Hummingbirds

DOMINIC J. WELLS* AND CHARLES P. ELLINGTON†

**Department of Veterinary Basic Sciences, Royal Veterinary College, London NWI OTU, United Kingdom; and †Department of Zoology, University of Cambridge, Cambridge, CB2 3EJ, United Kingdom*

I. Introduction
II. Metabolic Power Input
III. Muscle Power Output
IV. Maximal Muscle Power Output
V. Summary
References

I. Introduction

The preceding papers in this volume have dealt exclusively with aspects of exercise performance in vertebrate athletes. We conclude this volume by looking beyond the design constraints of the vertebrate plan and consider similarities and differences in exercise performance of an elite vertebrate athlete, the hovering hummingbird, and elite invertebrate athletes, hovering insect fliers.

Insects and hummingbirds exhibit extremes in design that can give insights into muscle function and limitations to exercise performance. The two groups of animals occupy similar ecological niches and show common adaptations for their roles, thereby providing a fascinating junction between the smallest vertebrate fliers and the upper range of insect aerialists.

True hovering fliers are found in three groups: the hummingbirds, the "primitive" insects, and the "advanced" insects. Hummingbird body masses range from 2 to 20 g and their wing lengths from 30 to 120 mm, and they operate at wingbeat frequencies of 15–80 Hz

(Greenewalt, 1962). Hummingbirds exhibit the highest mass-specific maximal oxygen consumptions recorded in vertebrates, on average 40–50 ml g^{-1} h^{-1} (Wells, 1990).

The group of insects that are most similar to the hummingbirds are the so-called "primitive" insects, which include dragonflies, grasshoppers, moths, and butterflies. These insects operate with synchronous muscles; that is, they contract in response to neural stimulation and have similar ultrastructural components to vertebrate muscle. The sphingid moths are of particular interest, as they span a size range that overlaps that of hummingbirds. In this group, body masses range from 0.35 to 3.5 g, wing lengths from 25 to 67 mm, and wingbeat frequencies from 25 to 66 Hz. There is a similarly high mass-specific oxygen consumption of 40–70 ml g^{-1} h^{-1} (Casey, 1981).

The "advanced" insects, for example, flies, beetles, bees, and bugs, operate their wings at high frequencies with stretch-activated asynchronous flight muscles. These muscles have a broadly similar sarcomeric organization compared to vertebrate muscle, but have little sarcoplasmic reticulum, because of reduced requirements for calcium pumping. These differences offer an interesting contrast to vertebrate muscle. Body masses of typical "advanced" insects range from 0.05 to 1.0 g, wing lengths from 7 to 24 mm, and wingbeat frequencies from 100 to 200 Hz. Mass-specific oxygen consumptions are on the order of 50–150 ml g^{-1} h^{-1} (Casey, 1989).

II. Metabolic Power Input

A plot of maximum recorded mass-specific oxygen consumption ($\dot{V}_{O_2max}$) against body mass clearly shows that fliers exceed the energetic requirements and/or aerobic capabilities of terrestrial animals (Fig. 1). The smallest fliers, insects and hummingbirds, employ hovering flight with very high metabolic rates; indeed, hovering is frequently cited as being the most energetically expensive form of locomotion (Weis-Fogh, 1972).

It is interesting to examine the structures responsible for the very high mass-specific oxygen consumption of hovering fliers (Fig. 2A). Skeletal muscle mitochondria are the site of utilization for the majority of the increased oxygen consumption associated with flight. Not surprisingly, hummingbird flight muscles have the highest mitochondrial volume densities recorded in vertebrate locomotor muscle (37.5% in Wells, 1990; 35% in Suarez *et al.*, 1991; 32–36% in Zerbinatti *et al.*, 1992). Interestingly, these muscles have mitochondrial volume

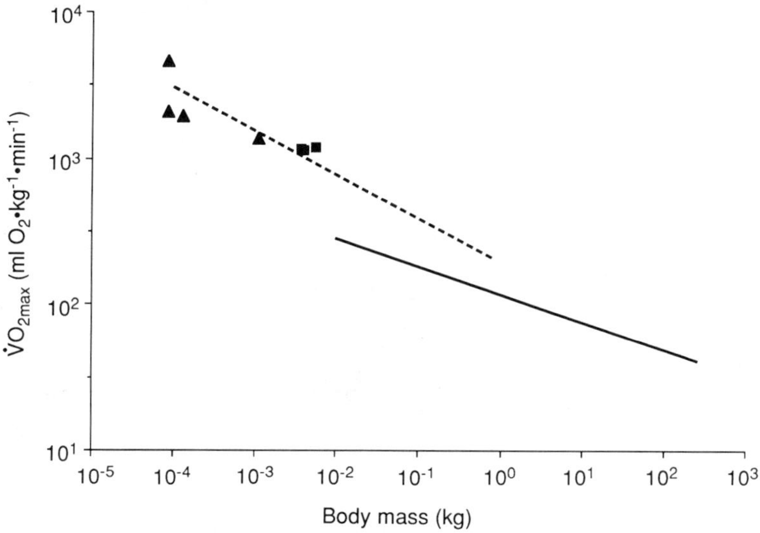

FIG. 1. Mass-specific maximal oxygen consumption ($\dot{V}_{O_2max}$) varies as a linear function of body mass among the most highly active (aerobic) invertebrates and vertebrates. However, the regression line for data obtained from flying insects, bats, and birds (dashed line; Berger, 1985; Carpenter, 1985; Casey, 1989; Casey *et al.*, 1985; Epting, 1980; Rothe *et al.*, 1987; Thomas, 1975, 1981; Tucker, 1966; Wells, 1990) shows a substantially higher mass-specific $\dot{V}_{O_2max}$ for fliers compared to runners (solid line; Taylor *et al.*, 1981). Triangle denotes insects and square denotes hummingbirds. Modified from Lindstedt *et al.* (1991).

densities that are equal to or higher than those found in hummingbird (Wells, 1990) and mammalian hearts (Hoppeler *et al.*, 1984). Over a large size range in the euglossine bees, there is a close relationship between decreasing body size and increasing mitochondrial volume density, which ranges from 31 to 43% (Casey and Ellington, 1989). A similar trend is seen in mammalian hearts, with increasing mitochondrial volume density as body size decreases (Hoppeler *et al.*, 1984). In contrast, human elite endurance athletes have an average mitochondrial volume density of 7.3% in one of the main locomotor muscles (Hoppeler *et al.*, 1973).

No endothermic vertebrate locomotor muscle has been described with a myofibrillar density less than 50% of the fiber volume. Because of the rapid contraction rates required of hovering flight, vertebrate and insect synchronous muscle fibers must contain a high density of sarcoplasmic reticulum (SR) for the rapid release and sequestration of calcium ions. Hence, with myofibrillar volume densities of 50–60% and SR occupying 10–20% of the fiber volume, mitochondrial densities

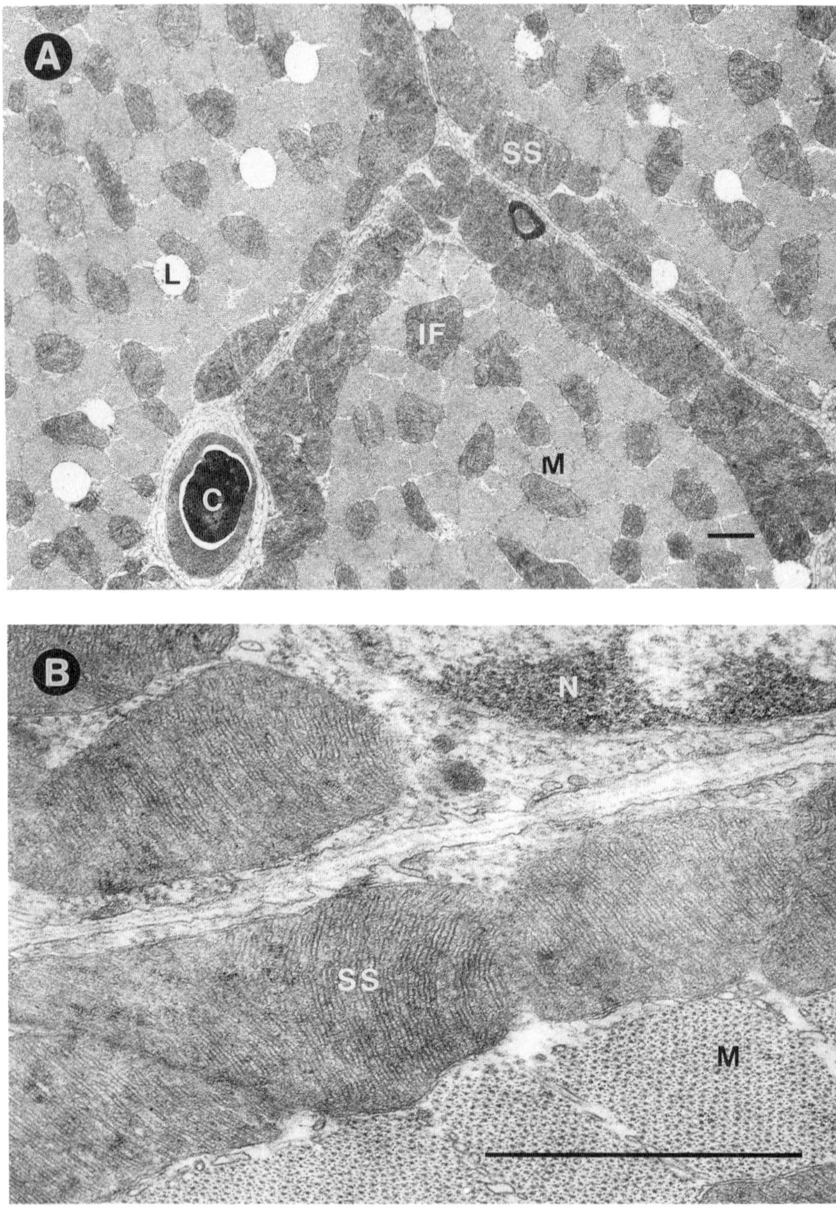

FIG. 2. (A) Electron micrograph of a transverse section of hummingbird *m. pectoralis* showing the high mitochondrial volume density. M, myofibrils; IF, intermyofibrillar mitochondria; SS, subsarcolemmal mitochondria; L, lipid droplets; C, capillary. The scale bar equals 1 μm. (B) Electron micrograph of a section of subsarcolemmal mitochondria from hummingbird *m. pectoralis* showing the dense packing of inner mitochondrial membranes. M, myofibrils; SS, subsarcolemmal mitochondria; N, myonucleus. The scale bar equals 1 μm.

cannot exceed 30–40%. Asynchronous muscle does not share this problem of vertebrate muscle, as there is little requirement for rapid calcium movement and hence little SR, enabling the muscle to have higher mitochondrial densities.

A notable feature of electron micrographs of skeletal muscle mitochondria from the hummingbirds is the dense packing of the inner mitochondrial membranes (Fig. 2B). Area estimates (surface densities) range from 58 to 77 m^2 cm^{-3} mitochondria (Suarez *et al.*, 1991; Wells, 1990; Zerbinatti *et al.*, 1992). These values exceed the range of packing densities found in mammalian locomotor muscles (20–40 m^2 cm^{-3}; Hoppeler and Lindstedt, 1985). In insect mitochondria the cristae are also densely packed, on the order of 50–60 m^2 cm^{-3} (Casey and Ellington, 1989). The close packing of the inner mitochondrial membranes in both the hummingbirds and insects can be seen as a means of increasing respiratory chain enzymes, and hence aerobic capacity, without unduly compromising the volume density of myofibrils. The similarities of packing densities also suggest that they may be at, or approaching, the limit for the Krebs cycle enzymes in the matrix to support the respiratory chain.

Given the high oxygen consumptions and high mitochondrial densities of hovering fliers, it is of interest to compare oxygen consumption per unit mitochondrial volume across different classes (Table I). What appears to be an invariant relationship in mammals (Hoppeler and Lindstedt, 1985) does not hold true for these other groups. Though the apparent disparity may, in part, be accounted for by differences in

TABLE I

COMPARISON OF VOLUME-SPECIFIC OXYGEN CONSUMPTION RATES OF WHOLE-BODY-AVERAGED SKELETAL MUSCLE MITOCHONDRIA IN MAMMALS, HUMMINGBIRDS, AND INSECTS (EUGLOSINE BEES) AT THE WHOLE-BODY MAXIMAL RATE OF OXYGEN CONSUMPTION ($\dot{V}_{O_2 max}$) OR WHOLE-BODY $\dot{V}_{O_2}$ DURING HOVERING FLIGHT ($\dot{V}_{O_2 hover}$)

Taxon	Metabolic intensity	Mitochondrial oxygen consumption (ml O_2 [STPD] min^{-1} cm^{-3})	Size dependence
Mammals	$\dot{V}_{O_2 max}$	5	No
Hummingbirds	$\dot{V}_{O_2 hover}$	8.7	?
	$\dot{V}_{O_2 max}$	13.5	?
Insets	$\dot{V}_{O_2 max}$	6.2–16	Yes

[a] Mammalian data from Hoppeler and Lindstedt (1985), hummingbird data from Wells (1990), and insect data from Casey and Ellington (1989).

inner mitochondrial membrane area, it cannot be fully explained in this manner.

It is possible that a better oxygen supply to the tissues is another factor accounting for the higher mitochondrial specific oxygen consumptions in hummingbirds and insects, compared to mammals. Hummingbirds possess lungs with almost 10-fold greater oxygen diffusing capacity than similar-sized mammals (Dubach, 1981). Their hearts are twice as large as would be predicted from allometric scaling (Schmidt-Nielsen, 1984), and their calculated cardiac outputs are phenomenal, approximately five times total body mass per minute (equivalent to a human heart pumping 350 liters min^{-1}). Additionally, Mathieu-Costello *et al.* (1992) have shown that the small-diameter muscle fibers in the hummingbird are extremely well endowed with capillaries. Insects have a simpler delivery system with only one convective and one diffusion step, in contrast to the two convective and two diffusion steps in vertebrates. Tracheoles carry oxygen directly to the muscles without the need for transport through the cardiovascular system. Hence, through altered mitochondrial morphology and a high delivery of oxygen to the locomotor muscles, hummingbirds and insects appear to have pushed aerobic metabolism as high as is structurally possible for locomotor muscles.

III. Muscle Power Output

So far we have considered the energy available for the locomotor muscles; the next question we must address is the capacity of the muscles to generate power output. The comparison between invertebrate and vertebrate fliers is facilitated by the recent collection of similar data sets for hummingbirds and bees. This allows direct comparison of the energetics and mechanics of flight. The analysis requires high-speed cinematography or video recordings *in vivo*, with a detailed *postmortem* analysis of muscle mass, wing mass, area, and shape. The mathematics for the calculation of specific components of the total muscle power output are described in Ellington (1984a–f). These components are: the inertial power required to oscillate the wing and associated air mass (virtual mass) during each wingbeat (P^*_{acc}), the induced power required to accelerate the air downward as a reaction to the lift force supporting the animal's mass (P^*_{ind}), the profile power required to overcome the frictional drag on the wing (P^*_{pro}), and the parasite power required to overcome body drag (P^*_{par}). This last component of power output is ignored in hovering flight because the body is not moving through the air. These different elements of the total power output from the flight muscles can be calculated from detailed kinematic and morphological analyses.

The contribution of these elements to total power output will depend on the fate of the kinetic energy released upon deceleration of the wing in the second phase of each half stroke. In the absence of elastic storage, this kinetic energy is lost, and the muscles must actively decelerate the wing. Hence

$$\text{Power output} = \frac{1}{2}\,(P^*_{\text{acc}} + P^*_{\text{ind}} + P^*_{\text{pro}})$$

as the metabolic cost of any negative work is negligible. For perfect elastic storage it is assumed that the kinetic energy is stored as elastic potential energy during deceleration and then released to accelerate the wing on the next stroke. If no energy is lost during storage, then the net inertial power requirements over the cycle is zero, and the power output is simply the sum of the induced and profile power requirements. Muscle power output can then be calculated from total power output by knowing the flight muscle mass. Overall muscle efficiency is the mechanical power output divided by the metabolic power input.

Looking first at insects (Table II), muscle mass-specific power output is largely independent of body mass for euglossine bees, spanning a range of body masses from 0.08 to 1.13 g (Casey and Ellington, 1989).

TABLE II

COMPARISON OF INSECT AND HUMMINGBIRD BODY-MASS-SPECIFIC METABOLIC POWER INPUT AND MUSCLE-MASS-SPECIFIC POWER OUTPUT DURING HOVERING FLIGHT[a]

		Muscle power output			
		Zero elastic storage		Perfect elastic storage	
Taxon	Metabolic power input ($W\ kg^{-1}$ body mass)	$W\ kg^{-1}$ muscle mass	% efficiency	$W\ kg^{-1}$ muscle mass	% efficiency
Insects[b]	200–642	200	8–34	100	4–16
Hummingbirds[c]	295	270	24.4	106	9.6

[a] Muscle power outputs and efficiencies are calculated for the extreme cases of zero elastic storage of energy and perfect storage of elastic energy.

[b] From Casey and Ellington (1989).

[c] From Wells (1993a).

A question remains as to which of the two extremes, zero or perfect elastic storage, most closely approaches the situation *in vivo*. Although it is tempting to choose the zero storage analysis, as these values are close to the theoretical maximal performance for muscle (Weis-Fogh and Alexander, 1977), a number of lines of evidence suggest this is unlikely. First, many insects can hover carrying an additional load of 50% of their body mass; under these conditions zero elastic storage would entail power outputs exceeding the theoretical maximum (see the following). Second, insects have structures that are well recognized as being suitable for elastic storage of kinetic energy (e.g., Alexander and Bennet-Clark, 1977; Ellington, 1984f).

Mass-specific muscle power output is independent of body size, but mass-specific oxygen consumption decreases with increasing body mass. Hence, efficiency must be lower in smaller animals. This holds true for either assumption about the degree of elastic storage, with efficiencies ranging from 8 to 34% for zero elastic storage and from 4 to 16% for perfect elastic storage (Casey and Ellington, 1989). It may seem counterintuitive that muscle efficiency should be lower in the case of high elastic storage, but in this situation the muscles will be doing less work for the same metabolic power input.

Similar values arise if we do the same calculations with data collected from hummingbirds (Table II). Once again, either we accept that hummingbirds operate at high efficiency with muscle power outputs equaling or exceeding the theoretical maximum, or we assume that they also exhibit elastic storage with a concomitant lower efficiency. Like insects, hummingbirds can hover while carrying loads, up to 30% of body mass experimentally, indicating that the animals are not operating at their maximum power output. The near constancy of wingbeat frequency during flight for a given individual suggests a tuned damped oscillator with an elastic restoring force (Greenewalt, 1960). Unfortunately, the hypothetical elastic structures in hummingbirds are not easily identified, as they are in insects.

If we take the perfect elastic storage analysis for hummingbird flight muscle, we calculate a muscle mass-specific power output of 106 W kg^{-1} at 9.6% efficiency. Is this figure reasonable? We can address this issue by considering the data for the efficiency of terrestrial locomotion. Efficiency has been shown by Heglund *et al.* (1982) to scale with body size; large animals like the horse operate at high efficiencies, and small animals like the mouse operate at low efficiencies (<10%). We have already described a similar phenomenon in insects. The calculated power output for a hovering hummingbird exhibiting good elastic storage fits well into the continuum between the mammal and insect data sets, and is consistent with the survey of Ellington (1991).

The scaling of efficiency has been related to the speed of muscle contraction, and certainly this holds true for the mammal and insect groups. Hummingbirds, with their high wingbeat frequencies, would therefore be predicted to operate at low efficiencies. An important biological question remains as to the reason for this correlation. It has been suggested that the requirement for rapid calcium pumping in the faster muscles is responsible for the scaling effect, but the asynchronous muscles of bees do not pump calcium at these high rates, arguing against that hypothesis. Any other hypothesis that fails to account for the similar relationship between muscle efficiency and body size seen in invertebrates and vertebrates must likewise provide only a partial answer.

Further evidence for the likely existence of elastic storage mechanisms in both insects and hummingbirds comes from *in vitro* analysis of power output, using work loop experiments to model the actual muscle mechanics in an oscillating system. Results to date using the work loop analysis demonstrate that muscle power outputs and efficiencies are low for both asynchronous and synchronous insect muscles (Table III). Hummingbird muscles are probably similar in performance to the synchronous muscles of the sphingid moth, *Manduca sexta*. This contention is supported by the excellent work of Biewener *et al.* (1992), who have directly measured the power output of the *pectoralis* muscle of European starlings *in vivo* during flight (104 W kg^{-1}). A calculated

TABLE III

Comparison of *in Vitro* Performances of Muscle from Several Insect and Vertebrate Species

Animal	Power output (W kg^{-1})	Efficiency (%)	Reference
Manduca sexta	100–120	9–11	Stevenson and Josephson (1990)
Schistocerca americana	75	6.4	Josephson and Stevenson (1989), Mizisin and Josephson (1987)
Insects with fibrillar muscle	33–76	—	Josephson (1985a,b)
Frog	35	—	Stevens (1988)
Fish	25–35	—	Altringham and Johnston (1990)
Rodent *diaphragmus*	44–50	—	Altringham and Young (1991)

power output of 100–110 W kg^{-1} for hummingbirds compares very favorably with the value they measured.

IV. Maximal Muscle Power Output

It appears that the hovering fliers are elite athletes in terms of their aerobic capacities, but probably less so in terms of the power output from their flight muscles. But perhaps we are maligning these creatures, insofar as we have only examined "normal" hovering flight. It has been commonly supposed that hovering flight, being so energetically expensive, is close to maximal performance. Recent data clearly show this not to be the case.

By attaching a small harness to hummingbirds, a sort of hummingbird *lederhosen*, lead weights can be added to increase the mass of the bird. This approach was used to simulate premigratory fattening. Experimental birds could be induced to hover while carrying an extra 30% of body mass, analogous to a human athlete running rapidly with a 20-kg backpack. The hummingbirds were able to increase metabolic power input and muscle power output to meet this demand (Wells, 1993b).

Insects could also be persuaded to carry substantial additional mass. This was done initially by attaching weights to bees, but it is easier to do so by taking advantage of their natural feeding patterns. *Ad libitum* ingestion of a nectar substitute can increase body mass by 50%. Gorged bees were able to fly immediately after feeding, again producing an impressive increase in muscle and metabolic performance (Ellington *et al.*, 1994).

Hummingbirds generated the extra lift required for loaded flight by increasing their wingbeat amplitude and maintaining a near-constant wingbeat frequency, reinforcing the concept of a tuned oscillating system with elastic storage of kinetic energy (Wells, 1993b). In contrast, insects increased the wingbeat frequency with a relatively fixed wingbeat amplitude (Ellington *et al.*, 1994). The difference in strategy probably reflects the need to operate the asynchronous flight muscles of the bees over small strains. The ability to change wingbeat amplitude clearly would be useful to hummingbirds, as it would allow them to exploit a wide range of food sources, foraging from both narrow and broad-shaped flowers.

The loading studies show that during normal hovering there is a substantial reserve performance capacity available in these animals, reiterating the contention that they are unlikely to operate without substantial elastic storage. In general, hummingbirds and insects seem to be able to lift approximately six times their flight muscle mass (Marden, 1987).

Hummingbirds, and possibly insects, must also increase their power output during horizontal flight at high speeds to overcome increased parasite drag, which increases in proportion to their speed cubed. This is clearly seen in metabolic studies of hummingbirds flying in a wind tunnel (Berger, 1985), but not in similar insect studies, probably because the insect wind tunnel did not allow sufficiently high speeds (Ellington *et al.*, 1990). In hummingbirds, the *pectoralis* is twice the size of the *supracoracoideus*, yet the muscles have equal roles in hovering flight. Probably at least half of the maximal power output of the *pectoralis* is unused during hovering flight. This reserve is used when the stroke plane is angled forward during fast horizontal flight, and hence we must be underestimating the maximal flight muscle power output when examining hovering animals.

Last but not least, we should recognize these fliers as being the supreme aerial acrobats, capable of almost unbelievably fast changes in direction. This ability is very difficult to quantify, but examples can be found in Greenewalt (1960) and Ellington (1984c).

V. Summary

Hummingbirds and insects are clearly extreme aerobic athletes. Hovering hummingbirds and insects exhibit the highest mass-specific metabolic rates found in vertebrates and invertebrates, respectively. Both groups of fliers have high mitochondrial volume densities in their locomotor muscles, but these do not exceed 35–40% of the fiber volume, presumably from a need to conserve myofibrils for force generation. A possible adaptation to this constraint is the observed greater packing of the inner mitochondrial membranes than occurs in mammalian mitochondria. Both hummingbirds and insects show higher rates of oxygen consumption per unit volume of mitochondria than do mammals. Additionally, volume-specific mitochondrial oxygen consumption in insects increases as body size decreases, unlike the size-independent pattern in mammals.

Aerodynamic analysis of power output during hovering flight strongly suggests that both insects and hummingbirds operate with considerable elastic storage of kinetic energy, thereby decreasing their inertial power requirements. Both groups appear to hover with muscle power output close to 100 W kg^{-1}. Muscle efficiency in hummingbirds is near

10%; in insects, muscle efficiency varies with body size, but at similarly low values. Scaling of efficiency with body size has also been reported in terrestrial mammals, suggesting a possible common mechanism. Both groups of hovering fliers can markedly increase their metabolic power inputs and mechanical power outputs above those required for basic hovering flight.

These elite aerial athletes offer considerable insight into the constraints and demands on animal design for maximal aerobic capacity. Additionally, the similarities shown between the different phyla suggest the existence of common mechanisms and limitations in metabolic and mechanical performance. Insects in particular offer a number of advantages in pursuing questions such as the cause of the allometric scaling of muscle efficiency; this scaling can be examined within families, genera, or species with the additional benefit that insect muscles also perform well *in vitro*.

References

Alexander, R. McN., and Bennet-Clark, H. C. (1977). Storage of elastic strain energy in muscle and other tissues. *Nature (London)* **265,** 114–117.

Altringham, J. D., and Johnston, I. A. (1990). Modelling muscle power output in a swimming fish. *J. Exp. Biol.* **148,** 395–402.

Altringham, J. D., and Young, I. A. (1991). Power output and the frequency of oscillatory work in mammalian diaphragm muscle: The effects of animal size. *J. Exp. Biol.* **157,** 381–389.

Berger, M. (1985). Sauerstoffverbrauch von kolibris (*Colibri coruscans* und *C. thalassinus*) beim horizontalflug. *BIONIA Rep.* **3,** 307–314.

Biewener, A. A., Dial, K. P., and Goslow, G. E. (1992). Pectoralis muscle force and power output during flight in the starling. *J. Exp. Biol.* **164,** 1–18.

Carpenter, R. E. (1985). Flight physiology of flying foxes, *Pteropus poliocephalus*. *J. Exp. Biol.* **114,** 619–647.

Casey, T. M. (1981). A comparison of mechanical and energetic estimates of flight cost for hovering sphinx moths. *J. Exp. Biol.* **91,** 117–129.

Casey, T. M. (1989). Oxygen consumption during flight. *In* "Insect Flight" (G. Goldsworthy and C. Wheeler, eds.), pp. 257–272. CRC Press, Boca Raton, FL.

Casey, T. M., and Ellington, C. P. (1989). Energetics of insect flight. *In* "Energy Transformations in Cells and Organisms" (W. Wieser and E. Gnaiger, eds.), pp. 200–210. Thieme, New York.

Casey, T. M., May, M. L., and Morgan, K. R. (1985). Flight energetics of euglossine bees in relation in morphology and wing stroke frequency. *J. Exp. Biol.* **116,** 271–289.

Dubach, M. (1981). Quantitative analysis of the respiratory system of the house sparrow, budgerigar and violet-eared hummingbird. *Respir. Physiol.* **46,** 43–60.

Ellington, C. P. (1984a). The aerodynamics of hovering insect flight. I. The quasi-steady analysis. *Philos. Trans. R. Soc. London, Ser. B* **305,** 1–15.

Ellington, C. P. (1984b). The aerodynamics of hovering insect flight. II. Morphological parameters. *Philos. Trans. R. Soc. London, Ser. B* **305,** 17–40.

Ellington, C. P. (1984c). The aerodynamics of hovering insect flight. III. Kinematics. *Philos. Trans. R. Soc. London, Ser. B* **305,** 41–78.

Ellington, C. P. (1984d). The aerodynamics of hovering insect flight. IV. Aerodynamic mechanisms. *Philos. Trans. R. Soc. London, Ser. B* **305,** 79–113.

Ellington, C. P. (1984e). The aerodynamics of hovering insect flight. V. A vortex theory. *Philos. Trans. R. Soc. London, Ser. B* **305,** 115–144.

Ellington, C. P. (1984f). The aerodynamics of hovering insect flight. VI. Lift and power requirements. *Philos. Trans. R. Soc. London, Ser. B* **305,** 145–181.

Ellington, C. P. (1991). Limitations on animal flight performance. *J. Exp. Biol.* **160,** 71–91.

Ellington, C. P., Machin, K. E., and Casey, T. M. (1990). Oxygen consumption of bumblebees in forward flight. *Nature (London)* **347,** 472–473.

Ellington, C. P. *et al.* (1994). In preparation.

Epting, R. J. (1980). Functional dependence of the power for hovering on wing disc loading in hummingbirds. *Physiol. Zool.* **53,** 347–357

Greenewalt, C. H. (1960). "Hummingbirds." Doubleday, New York.

Greenewalt, C. H. (1962). Dimensional relationships for flying animals. *Smithson. Misc. Collect.* **144,** 1–46.

Heglund, N. C., Fedak, M. A., Taylor, C. R., and Cavagna, G. A. (1982). Energetics and mechanics of terrestrial locomotion. IV. Total mechanical energy changes as a function of speed and body size in birds and mammals. *J. Exp. Biol.* **97,** 57–66.

Hoppeler, H., and Lindstedt, S. L. (1985). Malleability of skeletal muscle in overcoming limitations: Structural limitations. *J. Exp. Biol.* **115,** 355–364.

Hoppeler, H., Luthi, P., Claassen, H., Weibel, E. R., and Howald, H. (1973). The ultrastructure of the normal human skeletal muscle. *Pflüegers Arch.* **344,** 217–232.

Hoppeler, H., Lindstedt, S. L., Claassen, H., Taylor, C. R., Mattieu, O., and Weibel, E. R. (1984). Scaling mitochondrial volume in heart to body mass. *Respir. Physiol.* **55,** 131–137.

Josephson, R. K. (1985a). Mechanical power output from a striated muscle during cyclic contraction. *J. Exp. Biol.* **114,** 493–512.

Josephson, R. K. (1985b). The mechanical power output of tettigoniid wing muscle during singing and flight. *J. Exp. Biol.* **117,** 533–537.

Josephson, R. K., and Stevenson, R. D. (1989). The efficiency of an insect flight muscle. *Am. Zool.* **29,** 83A.

Lindstedt, S. L., Hokanson, J. F., Wells, D. J., Swain, S. D., Hoppeler, H., and Navarro, V. (1991). Running energetics in the pronghorn antelope. *Nature (London)* **353,** 748–750.

Marden, J. H. (1987). Maximum lift production during takeoff in flying animals. *J. Exp. Biol.* **130,** 235–258.

Mathieu-Costello, O., Suarez, R. K., and Hochachka, P. W. (1992). Capillary-to-fiber geometry and mitochondrial density in hummingbird flight muscle. *Respir. Physiol.* **89,** 113–132.

Mizisin, A. P., and Josephson, R. K. (1987). Mechanical power output of locust flight muscle. *J. Comp. Physiol. A* **160A,** 413–419.

Rothe, H.-J., Biesel, W., and Nachtigall, W. (1987). Pigeon flight in a wind tunnel. *J. Comp. Physiol. B* **157B,** 99–109.

Schmidt-Nielsen, K. (1984). "Scaling: Why Is Animal Size So Important?" Cambridge Univ. Press, Cambridge, UK.

Stevens, E. D. (1988). Effect of pH and stimulus phase on work done by isolated frog sartorius muscle during cyclical contraction. *J. Muscle Res. Cell Motil.* **9,** 329–333.

Stevenson, R. D., and Josephson, R. K. (1990). Effects of operating frequency and temperature on mechanical power output from moth flight muscle. *J. Exp. Biol.* **149,** 61–78.

Suarez, R. K., Lighton, J. R. B., Brown, G. S., and Mathieu-Costello, O. (1991). Mitochondrial respiration in hummingbird flight muscles. *Proc. Natl. Acad. Sci. U.S.A.* **88,** 4870–4873.

Taylor, C. R., Maloiy, G. M. O., Weibel, E. R., Langman, V. A., Kamau, J. M. Z., Seeherman, H. J., and Heglund, N. C. (1981). Design of the mammalian respiratory system. III. Scaling maximum aerobic capacity to body mass: Wild and domestic mammals. *Respir. Physiol.* **44,** 25–37.

Thomas, S. P. (1975). Metabolism during flight in two species of bats, *Phyllostomus hastatus* and *Pteropus gouldii. J. Exp. Biol.* **63,** 273–293.

Thomas, S. P. (1981). Ventilation and oxygen extraction in the bat *Pteropus gouldii* during rest and steady flight. *J. Exp. Biol.* **94,** 231–250.

Tucker, V. A. (1966). Oxygen consumption of a flying bird. *Science* **154,** 150–151.

Weis-Fogh, T. (1972). Energetics of hovering flight in hummingbirds and *Drosophila. J. Exp. Biol.* **56,** 79–104.

Weis-Fogh, T., and Alexander, R. McN. (1977). The sustained power output from striated muscle. *In* "Scale Effects in Animal Locomotion" (T. J. Pedley, ed.), pp. 511–525. Academic Press, London.

Wells, D. J. (1990). Hummingbird flight physiology: Muscle performance and ecological constraints. Ph.D. Dissertation, University of Wyoming, Laramie.

Wells, D. J. (1993a). Muscle performance in hovering hummingbirds. *J. Exp. Biol.* **178,** 39–57.

Wells, D. J. (1993b). Ecological correlates of hovering hummingbird flight. *J. Exp. Biol.* **178,** 59–70.

Zerbinatti, C. V., Bicudo, J. E. P. W., and Lindstedt, S. L. (1992). Effects of body mass (M_b) on mitochondrial and capillary volume densities, and mitochondrial inner surface area Sv(im,mi) in flight muscle, heart and liver of hummingbirds. *Physiologist* **35,** 234.

Index

A

Acceleration, in fish, 30–33
Adductor magnus, anuran jumping performance, 75
Aerobic speed, maximal, reptile locomotion effects, 123–124
Agnathans, *see also* Fish
  swimming gait, 17–18
Allometry
  anuran muscles
    interspecific, 76–78
    intraspecific, 82–83
    scaling predictions, 83–85
  mammalian oxygen transport system, 211
Angles, takeoff, and anuran jumping performance, 56–57, 61
Anurans, *see also* Frogs; toads
  allometry, proportionalities, 84
  jumping ability
    interspecific variation, 79–81
    mechanics, 52–55
    muscles, 74–75
  muscle kinetics, 68–69
  thermoregulation effects, 90–93
ATP (adenosine 5′-triphosphate), in reptilian activity bursts, 128–130

B

Birds, *see also specific types*
  exercise, energy demands, 141–143
  cardiovascular specialization, 154–162
  conditioning and deconditioning, 176–179
  diving, physiological adaptations, 172–175
  locomotion patterns, comparison with reptiles, 120–121
  prolonged flights
    high altitude provisions, 150–154
    physiological adaptations, 144–148
    thermoregulation and water conservation, 168–171
    tissue specializations, 162–164
  pulmonary specialization, 144–154
Blood
  birds, properties, 158–162
  mammals, hemoglobin oxygen kinetics, 206–207
Blood–gas barrier, 145–148
Burst speed exertion, in reptiles
  ecological and evolutionary purposes, 131–132
  metabolic support, 127–130

C

Cardiovascular system, avian, 154–158

E

Electromyography
  semimembranosus muscle, 100–102
  anuran hindlimb muscles, 75–76
EMG, *see* Electromyography
Endotherms, gait speed, 18
Endurance activity, reptile locomotion systems, 124–127
Endurance–speed curve, factors affecting, 14–16
Energetic performance
  definition, 5
  swimming levels, 12–13
Exertion, burst speed, *see* Burst speed exertion

F

Fast-starts, fish, 31–33
Fish, *see also* Agnathans
  evolution, 2

flow refuging types, 20–22
gait evolution, 17–18
locomotion performance, 1–2
comparison with reptiles, 120–121
muscle types, 6–10
propulsion types, 2
swimming
acceleration types, 30–32
mechanics, 3–5
zero ground speed, 20
turning circles, 40
Fliers, hovering
muscle power output differences, 224–227
types, 219–220
Flight, bird, energy demands, 141
Flow refuging, fish types, 20–22
Formulas, anuran jumping performance, 52–55
Frogs
jumping
body-mass-specific work output, 87–90
performance, muscle activity and shortening, 74–75, 98–102
muscles, contractile properties, 63–67

G

Gaits, fish, species variation, 18–19
Gas exchange system, crosscurrent, prolonged flights of birds, 144–145
Glycolysis, anaerobic, burst speed exertion in reptiles, 127–130
Gray's Paradox, 2
Ground reaction forces, and anuran jumping performance, 58–60, 93–95

H

Hummingbirds
elastic storage mechanisms, 227–228
metabolic power input, mitochondrial density, 221–224
muscle power output, 224–227

I

Insects
elastic storage mechanisms, 227–228
metabolic power input, 220–224
metabolic rates, 229–230
mitochondrial density levels, 221–223

J

Jumping, anuran
body-mass-specific work output, 87–90
detailed studies, 93–98
formulae for distance, 52–55
mechanical work and power calculations, 60–63
muscle contractile properties, 63–67
muscles utilized, 74–75
takeoff angle effects, 56–57

K

Kinematics, jumping performance measurements, 95–98
Kinetics, anuran muscle, 68–69

L

Lactic acid, reptile locomotion inhibition, 127–128
Locomotion systems,
birds
exercise energy demands, 141–144
mitochondrial density, 221–224
fish, components, 7–10
mammals, fundamental elements, 200–202

M

Mammals
body size and aerobic performance, 193–196, 210–212

locomotion
comparison with reptiles, 120–121
fundamental elements, 200–202
muscles, oxygen demands, 196–198
Metabolism, anaerobic
hummingbirds, 229–230
reptile locomotion system, 130–131
Muscles
aerobic performance, mitochondrial content effects, 167
anurans, interspecific allometry, 76–78
birds, seasonal changes, 176–177
fish, 7–10
frog, contractile properties, 63–67
hummingbirds, mitochondrial density levels, 221–223
mammals
aerobic oxygen demands, 196–198
oxygen delivery, 203–204
Myoglobin, birds, 163
seasonal changes, 179

O

Oxygen consumption, maximal
birds, 141–143
hummingbirds, 220
reptiles, 117
Oxygen transport systems
birds, pulmonary specializations, 144–145
mammals
body size effects, 195
pulmonary ventilation, 205–207

P

Penguins, diving adaptations
pulmonary specializations, 172–174
Power output, maximum, anurans
muscle isotonic contractions, 72–74
Propulsion, fish
MPF
versus BCF, definition, 2
limitations of, 22
multigear system, 10–12
types, 8

R

Reptiles, *see also* Snakes
endotherm transformation, 114
endurance activity, 116–118
ecological and evolutionary implications, 124–127
locomotion patterns
comparison with endotherms, 120–121
energy expenditures, 118–121
locomotion structures, 115–116
Respiratory system, avian, 144–154

S

Sarcoplasmic reticulum, in synchronous and asynchronous muscle, 221–223
Semimembranosus muscle, EMG activity and shortening, 100–102
Snakes, *see also* Reptiles
locomotion patterns, 116
Speed, zero ground, 20, 120
Swimming
birds, metabolic rate studies, 143
fish
acceleration types, 30–32
endurance curves, 12–15
intermittent, 10–12, 30
mechanics of, 3–5
performance limitations, 20–24
sprints, 28–30

T

Thermoregulation
anurans, 90–93
effect on jumping performance, 68–72, 80–82
prolonged bird flights, effects on, 168
Toads, *see also* Anurans
jumping performance, 88–90

V

Ventilation, prolonged flights of birds, 148–150

W

Water conservation, in birds
physiological adaptations, 171

Z

Zero ground speed, and fish, locomotion
energy expenditures, 20, 120

ISBN 0-12-039239-9

90018

9 780120 392391